NEW VANGUARD 309

THE HAWK AIR DEFENSE MISSILE SYSTEM

**MARC ROMANYCH &
JACQUELINE SCOTT**

ILLUSTRATED BY
IRENE CANO RODRÍGUEZ

OSPREY PUBLISHING
Bloomsbury Publishing Plc
Kemp House, Chawley Park, Cumnor Hill, Oxford OX2 9PH, UK
29 Earlsfort Terrace, Dublin 2, Ireland
1385 Broadway, 5th Floor, New York, NY 10018, USA
E-mail: info@ospreypublishing.com
www.ospreypublishing.com

OSPREY is a trademark of Osprey Publishing Ltd

First published in Great Britain in 2022

A catalogue record for this book is available from the British Library.

ISBN: PB 9781472852212; eBook: 9781472852182;
ePDF: 9781472852199; XML: 9781472852175

22 23 24 25 26 10 9 8 7 6 5 4 3 2 1

Index by Alison Worthington
Typeset by PDQ Digital Media Solutions, Bungay, UK
Printed and bound in India by Replika Press Private Ltd.

Osprey Publishing supports the Woodland Trust, the UK's leading woodland
conservation charity.

To find out more about our authors and books visit
www.ospreypublishing.com. Here you will find extracts, author
interviews, details of forthcoming events and the option to sign up for our
newsletter.

ACKNOWLEDGEMENTS

The authors wish to acknowledge the contributions of HAWK missile
system veterans who shared their knowledge and experiences with us as
we researched and wrote this book. Especially helpful were Colonel
(USA Retired) Michael R. Gonzales, who served in 6-52 ADA and provided
information about US Army National Guard HAWK battalions; Chief Warrant
Officer 4 (USA Retired) Roger M. Scholbe, who served for 23 years in various
HAWK assignments, including nine HAWK battalions, and shared his
encyclopedic knowledge of the HAWK system; Sergeant First Class
(USA Retired) Joseph Marcheggiani, who served in three HAWK battalions
and contributed his first-hand knowledge of HAWK systems and unit
operations; Calvin Bridges, who served in 3-7 ADA during its time as a
self-propelled HAWK unit and introduced us to numerous veterans who
shared their experiences; Oberfeldwebel a.D. Ulrich Kroemer, who served in
1./FlaRakBtl 36 and provided hard-to-find information about European
HAWK units; and James Crabtree, who served as a tactical control officer in
B Btry, 2-1 ADA during the Gulf War and offered key insights into the unit's
operations.

CONTENTS

THE HAWK AIR DEFENSE MISSILE SYSTEM

INTRODUCTION

HAWK (Homing All the Way Killer) was the world's first mobile surface-to-air missile system, and during the Cold War, one of the world's most sophisticated and widely-deployed medium-range air defense weapons. Designed to counter the threat posed by Soviet jet aircraft of the 1950s, the HAWK system became operational in 1959, just two years before the Berlin Crisis. At its peak, HAWK saw front-line service with the US military in the Far East, Panama, Europe, and the Middle East and more than 25 allied nations worldwide. The US also deployed HAWK units in combat roles during the Cuban Missile Crisis, Vietnam War, and Persian Gulf War to protect ground forces and vital installations.

The HAWK system was respected by friendly and opposing air forces alike for its lethality and proved itself in combat while serving with several foreign nations. During the Arab–Israeli Wars, Iran–Iraq War, Chadian–Libyan conflict, and Iraqi invasion of Kuwait, HAWK missiles were credited

To intercept high-speed, maneuverable aircraft, the HAWK missile had to be fast from the moment it left the launcher. Depending on the model, missiles flew at speeds between Mach 2.0 and 2.5. This missile was fired by C Btry, 3-68 ADA in 1983 at Onslow Beach, Camp Lejeune, North Carolina during a live-fire exercise. (M. Romanych)

Several prototype pieces of HAWK equipment were later significantly redesigned to improve their utility in the field. This is Raytheon's prototype Continuous Wave Acquisition Radar. (M. Romanych)

with shooting down approximately 100 combat aircraft. So great was Soviet concern about HAWK's lethality that it developed electronic jammers, anti-radiation missiles, and special tactics to attack the HAWK system. After the Berlin Wall fell in 1989 and the threat of Soviet air attack diminished, the US military retired its HAWK systems, though several foreign nations continue to use HAWK well into the 21st century.

DESIGN AND DEVELOPMENT

Development of the HAWK missile system began in 1952 to fill the US Army's need for a mobile air defense weapon that could engage low-flying jets, such as the Soviet MiG-15 which was encountered by the US Air Force during the Korean War. The project, initially called SAM-A-18 and then later HAWK, featured extensive use of continuous-wave radar technology. Whereas pulse-type air defense radars were hampered by ground clutter (i.e., radar signals reflected off terrain and vegetation that obscure real, or even create false, radar targets), HAWK's continuous-wave radars and semi-active missile guidance system were designed to ignore ground clutter by using the Doppler effect to distinguish between stationary objects and moving targets.

In 1954, the Army selected Raytheon, an electronics and radar manufacturer that pioneered both CW radars and semi-active homing missile guidance, to design and build the HAWK missile system. Raytheon built the system in an accelerated program, with the missile being the first part of the system to be engineered. Only two years after the contract was awarded, the missile was successfully test launched at White Sands Proving Ground in New Mexico. In subsequent tests, missiles intercepted maneuvering drones flown in single and multiple target formations. The Army briefly considered modifying the missile to accommodate a nuclear warhead for

HAWK radars, such as these Basic HAWK Pulse Acquisition Radars, were manufactured at Raytheon's plant in Andover, Massachusetts. Missile components were produced in nearby Waltham, and launchers were built at Northrop's facility in Anaheim, California. (M. Romanych)

destroying low-flying nuclear-laden bombers, but after test launching a prototype missile carrying a dummy warhead, the project was cancelled, and HAWK was not nuclear-armed. A complete prototype HAWK system was completed in July 1957, and in November that year, $41 million was allocated to procure ten battery sets of equipment and 176 missiles. The next year, the prototype system successfully engaged a drone flying at 410ft (125m) above ground. In subsequent tests, the system destroyed an Honest John surface-to-surface rocket in the first-ever intercept of a rocket by a missile, then successfully intercepted a smaller Little John rocket and a Corporal tactical ballistic missile.

The HAWK system was designed to engage aircraft flying between ground level and 50,000ft (15,000m) at speeds up to Mach 2, with a maximum engagement range of 32km (20 miles). To counter electronic jamming, electronic counter-countermeasures (ECCM) were built into the radars and missiles. These capabilities were considered sufficient for engaging Soviet supersonic aircraft, particularly the then-new Soviet MiG-21 and Su-7 aircraft. One of HAWK's key features was its mobility. All major pieces of equipment were mounted on either trailers or vehicles for rapid tactical movement and small enough for easy transport by aircraft or helicopter. In contrast, the Soviets' equivalent surface-to-air system, the S-125

The Pulse Acquisition Radar detected medium altitude targets. This AN/MPQ-50 radar of A Btry, 2-51 ADA at Fort Riley is undergoing maintenance checks prior to a field exercise in 1983. Cages for transporting sections of the radar's antenna are loaded in the bed of the 2½-ton cargo truck. (NARA)

(NATO name: SA-3 GOA), had a 25km (15 mile) range, possessed limited low altitude engagement and ECCM capabilities, and was only semi-mobile.

Raytheon began producing HAWK equipment in 1959. After manufacture, the equipment was delivered to the US Army at Fort Bliss and the missiles were sent to the Red River Arsenal, both in Texas. HAWK became operational in 1959 with the activation of the Army's first HAWK missile battalion at Fort Bliss. That same year, an agreement was made with NATO allies France, Italy, Netherlands, Belgium, and West Germany for co-production of HAWK components in Europe. During three decades of service with the US military and more than six decades with other militaries, the HAWK system underwent numerous modifications and several significant upgrades. When subsequent versions of the system were developed, the initial version that was first produced and fielded in 1959 became known as Basic HAWK.

Basic HAWK – the 1960s

The HAWK missile system had four major design components: radars, missile launchers, missiles, and fire control equipment. Basic HAWK's two acquisition radars were the AN/MPQ-35 Pulse Acquisition Radar (PAR) for long-range detection of medium altitude aircraft and the AN/MPQ-34 Continuous Wave Acquisition Radar (CWAR) for detecting aircraft flying close to the ground. The PAR was a conventional pulse-type radar with a large convex open lattice antenna array, while the CWAR was a continuous-wave radar that used the Doppler effect to distinguish low-flying targets from ground clutter. Smaller than the PAR, the CWAR had a flat-faced, rectangular-shaped antenna. When in operation, the PAR's and CWAR's antennas rotated in synchronization with each other at 20 rotations per minute to provide 360-degree radar coverage. The PAR's maximum detection range was approximately 100km (62 miles), while the CWAR's range was about 40km (25 miles). Built into the PAR was an Identification Friend or Foe (IFF) system, which was a radio transponder that could differentiate enemy from friendly aircraft by exchanging encrypted data signals with friendly aircraft equipped with compatible IFF systems.

Two continuous-wave tracking radars (also called illuminators), either the AN/MPQ-33 Low Power Illuminator Radar (LOPIR) or AN/MPQ-39 High Power Illuminator Radar (HIPIR), were used to track, or "illuminate," targets during missile intercept. The first HAWK systems were equipped with LOPIRs, but after a few years, were replaced by more powerful HIPIRs that could track aircraft at longer range – up to 60km (37 miles) and were less susceptible to electronic jamming. Both models of tracking radars had two large side-by-side dish-type antennas (one transmitter and one receiver), giving them a distinctive, if not odd, appearance. Unlike the acquisition radars, the tracking radars' antennas did not constantly rotate, but rather, moved left or right in azimuth and up and down

To detect low-flying aircraft, the HAWK system had the Continuous Wave Acquisition Radar. This Basic HAWK AN/MPQ-34 was operated by C Btry, 6-71 Arty at Cam Ranh Bay, Vietnam in 1965. (NARA)

in elevation in a box search to find and illuminate targets. With two tracking radars per system, HAWK could engage two targets simultaneously. Both the LOPIR and HIPIR relied on the Doppler effect to track targets. During target engagement, if enemy electronic jamming prevented a tracking radar from determining range to its target, then the HAWK system employed an AN/MPQ-37 Range-Only Radar (ROR) as an alternate means to get the range information needed by the tracking radar. The ROR's parabolic antenna resembled a small satellite dish receiver and was difficult to jam because of its different wave form (pulse rather than continuous wave) and radiating frequency.

The HAWK system had six XM78E3 missile launchers, each armed with three missiles. Launchers had wheels and were equipped with outriggers to provide a stable and level platform for missile launch. The upper part of a launcher – the boom – could rotate 360-degrees, permitting missiles to fire in any direction. Three launchers were connected to each tracking radar via a Launching Section Control Box (commonly referred to by the crews as a Junction- or J-Box). The J-Box was a rectangular metal case containing electronics to route data between the tracking radar and its three launchers and control the launchers during missile firing, reloading, or maintenance. A tracking radar with J-Box and three launchers comprised one firing section. Each of the HAWK system's two firing sections could engage one target at a time.

HAWK's supersonic MIM-23A missile was powered by a solid rocket propulsion system. The missile had a semi-active homing guidance system and large triangular fins with elevons to control direction of flight. Armed with a high-explosive fragmentation warhead, missiles detonated by either impact or proximity fuses. If necessary, missiles could also be command-detonated by the crew. The HAWK system's basic load was 36 missiles; 18 loaded on launchers and another 18 stored on truck- or trailer-mounted missile storage pallets. Each pallet held three missiles. Two self-propelled tracked XM501E2 loader-transporters (one per firing section) were used to transfer missiles between launchers and missile pallets.

The primary fire control unit was the AN/TSW-2 Battery Control Central (BCC). This large rectangular shelter was mounted on the bed of a

---

A

HOW THE BASIC HAWK SYSTEM WORKED

The HAWK missile system had 360-degree target surveillance and missile launch capability. Depending on an aircraft's size, speed, direction, and altitude, HAWK could detect medium altitude aircraft (1) up to 100km (62 miles) with its Pulse Acquisition Radar (2) and low-flying aircraft up to 60km (37 miles) with the Continuous Wave Acquisition Radar (3). Once detected, aircraft location information from the radars was displayed in the Battery Control Central (4), where the crew identified aircraft detected by the acquisition radars as either hostile or friendly. If an aircraft was identified as hostile, it was designated as a target and assigned to a tracking radar (5) for engagement. The tracking radar then "illuminated" the target with electromagnetic energy and tracked the target's location. If, because of enemy electronic jamming, the tracking radar was unable to determine range to the target, then the Range-Only Radar (6) was used as an alternate means to get the needed range information. Once the target was close enough for engagement – 32km (20 miles) or less – a missile was launched (7). Two missiles were launched if the target was closer than 20km (12 miles). After launch, the missile's onboard semi-active CW radar homed in on the target by following electromagnetic energy reflected off the target by the tracking radar. Upon intercept with the target, the missile detonated and destroyed the enemy aircraft (8). A typical engagement sequence took about 30 seconds. With its two tracking radars, the HAWK system could engage two targets simultaneously and rapidly engage multiple targets.

KEY

Components of the Basic HAWK System:

A. Pulse Acquisition Radar
B. Continuous Wave Acquisition Radar
C. Range-Only Radar
D. High Power Illuminator Radar

E. Battery Control Central
F. Launching Section Control Box
G. Missile Launcher

Note: Layout of the HAWK system is not to scale, and the system's diesel generators and cables are not shown.

M36 2½-ton cargo truck. The BCC contained five consoles with radarscopes, status displays, and the controls used by the crew to operate the HAWK system during tactical operations. The BCC was manned by five personnel who worked together to identify, select, and engage targets. The Basic HAWK system also had secondary fire control equipment, the AN/TSW-4 Assault Fire Command Console (AFCC). Operated by two crew members, the AFCC was a simplified, compact version of the BCC, housed in a man-portable 450lb equipment case. The AFCC contained a radar display and the electronic controls needed to operate a CWAR and a firing section (i.e., a tracking radar with three launchers). This "stripped-down" part of the HAWK system, called an Assault Fire Unit (AFU), was designed to operate independently of the BCC and the rest of the HAWK system. Detaching the AFU and deploying it away from the rest of the HAWK system gave a firing battery the ability to extend air defense coverage over a larger area or keep one firing section operational while the other moved. The AFU was one of Basic HAWK's features that grew in importance in subsequent versions of the system until it eventually became the system's primary operating configuration in the late 1980s.

A complete HAWK system was inter-connected by 17 375ft-long (114 metres) data cables that passed target and firing command information between the radars, fire control equipment, J-Box, and launchers. Twelve power cables, similar in size and length to the data cables, connected the

The first tracking radar produced for the HAWK system was the AN/MPQ-33 Low Power Illuminator Radar. Its distinctive circular structures were the radar's receiver (left) and transmitter (right) antennas. A soldier is checking the electronic chassis that calculated the data the missile needed to intercept a target. This radar belonged to A Btry, 2-57 Arty near Ansbach, Germany in 1961. (NARA)

The High Power Illuminator Radar replaced the Basic HAWK system's Low Power Illuminator in the mid-1960s. This AN/MPQ-46 of the 2d LAAM Bn is undergoing system checks during a training exercise in Hawaii in 1983. Underneath and beside the radar are a Launching Section Control Box (J-Box), several cable reels, and a mechanic's toolbox. (NARA)

The Range-Only Radar was a specialized pulse radar used to determine a target's range when the tracking radars were electronically jammed. The radar emitted a narrow beam on a different frequency band than the other radars. This AN/MPQ-51 was photographed during field testing of the IHAWK system at Fort Bliss in 1969. (NARA)

radars, launchers, and BCC to trailer-mounted 45kw (kilowatt) diesel generators. Five generators were needed to power HAWK systems equipped with LOPIRs, and seven generators powered systems with HIPIRs. Cable length was important because it determined the maximum distance the system's components could be emplaced from each other. To move a complete HAWK system with its missiles and associated equipment required 23 2½-ton M35 and M36 cargo trucks, while an AFU needed just six trucks.

A Basic HAWK XM78E3 launcher at White Sands Missile Range, New Mexico in 1960. Each launcher carried three missiles. The first HAWK missiles were painted white with black wings, but by the mid-1960s missiles were painted olive drab. (NARA)

Once deployed, the Basic HAWK system suffered reliability problems because its vacuum tube electronics were ill-suited to the rigors of 24-hour field operations. Particularly troublesome were the electronics of the BCC and radars, which were easily damaged by fluctuations in the electricity from the diesel generators. Performance limitations became apparent as well. The CWAR and tracking radars had difficulties differentiating slow, low-flying targets from ground clutter, and the system's estimated single-shot kill probability was low (about 60 percent), which meant two missiles had be fired at a target to ensure destruction. Additionally, the system's engagement reaction time of 30 seconds was too slow to respond to faster aircraft. As a result, several programs were initiated to modernize Basic HAWK and improve its effectiveness against the most modern Soviet combat aircraft such as the Su-15 fighter and Tu-22 bomber.

Self-Propelled HAWK – the early 1970s

In 1965, the US Army contracted Raytheon to enhance HAWK's utility as a forward area air defense system. Raytheon designed two new pieces of equipment for the AFU, the XM727 Self-Propelled Launcher and

AN/MSW-9 Platoon Command Post (PCP). The self-propelled launcher consisted of a towed launcher (minus its wheels, outriggers, and a few other parts) mounted to a modified tracked M548 cargo carrier. The launcher was further altered to carry missiles on the move, eliminating the need to transfer missiles to and from the launcher. This tracked carrier had a mechanism for self-leveling the launcher, an onboard 60kw diesel power generator, and electric-powered cable reels to rapidly lay and rewind power and data cables. The sum of these changes reduced the number of vehicles needed by the AFU and significantly shortened the time needed to march-order and emplace a launcher section. More than 70 self-propelled launchers, each costing approximately $200,000, were built. The PCP was a trailer-mounted equipment shelter containing the AFCC, a new IFF system, and radios for tactical communications. Prototype equipment was delivered in early 1967, and after several design changes, was put into production and fielded in 1969.

While the self-propelled launchers may have looked good on paper, they did not perform well in practice. Field operations revealed the missile's electronics were too sensitive for transport on a tracked carrier. Launchers had frequent mechanical problems, in particular oil and hydraulic leaks that occasionally sparked onboard fires, including at least one instance of a launcher catching fire while carrying missiles. After several years in the field, the self-propelled launchers were deemed a failure and replaced with towed launchers when Basic HAWK was upgraded to Improved HAWK. On the other hand, the addition of the PCP was successful and remained part of the HAWK system for the rest of its service life. Raytheon and Lockheed revisited the concept of improving HAWK's mobility in the late 1970s when they developed an articulated eight-wheel-drive, heavy duty "Dragon Wagon" vehicle to carry the AFU's radars, PCP, and launchers. A prototype was built and tested, but never produced.

HAWK equipment nomenclatures					
Radar	Basic HAWK	IHAWK	PIP I	PIP II	PIP III
PAR	AN/MPQ-35	AN/MPQ-50			
CWAR	AN/MPQ-34	AN/MPQ-48	AN/MPQ-55		AN/MPQ-62
LOPIR	AN/MPQ-33	None			
HIPIR	AN/MPQ-39	AN/MPQ-46		AN/MPQ-57	AN/MPQ-61
ROR	AN/MPQ-37	AN/MPQ-51			None
BCC	AN/TSW-2	AN/TSW-8	AN/TSW-11		None
AFCC	AN/TSW-4	None	None	None	None
PCP	AN/MSW-9 (SP HAWK)	AN/MSW-11		AN/MSW-13 (Army) AN/MSW-14 (USMC)	AN/MSW-20
BCP	None				AN/MSW-21
ICC	None	AN/MSQ-95		AN/MSQ-110 (Army) AN/MSQ-111 (USMC)	None
Towed Launcher	XM78E3	XM192E1			
SP Launcher	XM727	None			
Loader-transporter	XM501E2	XM501E3			
Missile	MIM-23A	MIM-23B		MIM-23C/D	MIM-23E/F

Improved HAWK – the 1970s

During development of the self-propelled HAWK, the US Army also initiated a program to modernize the Basic HAWK system. The program – called Improved HAWK (IHAWK) – began in 1966, with equipment production commencing three years later. While IHAWK's design principles, system configuration, and outward appearance were the same as Basic HAWK, electronically, IHAWK was vastly different with many, but not all, electronic components upgraded from vacuum tube to solid-state. These upgrades significantly improved the system's reliability, target detection in ground clutter, performance against electronic jamming, and ability to engage faster and more maneuverable aircraft. To differentiate Improved from Basic HAWK, the nomenclature of IHAWK's major equipment items was often prefixed with "Improved" (or just "I"), for example IPAR, ICWAR, IBCC, etc. This distinction was dropped when IHAWK received its first major upgrade in the late 1970s.

Tracked loader-transporters were used to transfer missiles from the missile storage pallets to the launchers. This XM501E2 loader-transporter is on the tactical site of A Btry, 3-7 Arty near Schweinfurt, Germany in 1960. In the background is the site's missile storage area. (NARA)

IHAWK's radars were the AN/MPQ-50 PAR, AN/MPQ-48 CWAR, AN/MPQ-46 HIPIR, and AN/MPQ-51 ROR. Along with enhanced target detection, these radars had greater range. The PAR's range was extended to 120km (75 miles), the CWAR's to 60km (37 miles), and the HIPIR's tracking capability to 80km (50 miles). The ROR's performance did not significantly change. Basic HAWK's IFF system was removed from the PAR and replaced by a new system mounted to the fire control vans. Because the new radars required more electric power, larger 60kw generators replaced the 45kw generators.

The fire control units were upgraded too. The self-propelled HAWK's PCP was re-engineered into a new and far more capable AN/MSW-11 PCP. The old PCP's AFCC was replaced by a much easier to use radar console and a solid-state computer that reduced system engagement time by automating target detection and selection. Operated by a crew of four, the PCP now housed new IFF and communications systems. A new piece of equipment, the AN/MSQ-95 Information Coordination Central (ICC), was added to the IHAWK system to give the BCC the same data processing, IFF, and communications capabilities as the PCP. The ICC was similar to the PCP but did not have a radar console. The ICC operated in conjunction with the AN/TSW-8 BCC and was connected to it via seven 15ft data cables, while the BCC was partially converted to solid-state circuitry to permit information from the ICC's computer to be displayed on its radarscopes.

IHAWK's new missile, designated the MIM-23B, possessed a more powerful rocket motor that extended its range to 40km (25 miles) and altitude to 18,000m (59,000ft). A larger, more lethal warhead and a guidance package improved performance against fast-moving low altitude aircraft. Additionally, the MIM-23B was a "certified round," meaning it did not require significant maintenance by unit personnel. To accommodate

The heart of the HAWK system was the fire control units from which the crew monitored and engaged targets. This truck-mounted Battery Control Central of the 3d LAAM Bn is emplaced on Onslow Beach, Camp Lejeune, North Carolina in 1986 during a unit live-fire exercise. Behind it is an Information Control Central with its roof-mounted Identification Friend or Foe antenna. (NARA)

the new missile, the launchers and loader-transporters underwent minor modifications.

The first IHAWK battalion was equipped in 1972 at Fort Bliss, and by 1978 all US Army and Marine Corps HAWK batteries were upgraded. Although IHAWK's reliability was much better than Basic HAWK, the system still required intensive maintenance and a constant supply of repair parts. IHAWK's performance was superior too, with an 85 percent probability of kill (as proved by Israel during the October 1973 War). However, by the late 1970s, IHAWK's ability to successfully engage new Soviet high-performance MiG-27s, Su-24s, and Su-25s, many of which were equipped with electronic warfare equipment, became uncertain. Furthermore, intelligence revealed the Soviets were fielding an Mi-8SMV airborne jammer helicopter (NATO: Hip J) and anti-radiation missiles to attack IHAWK radars.

The HAWK system's two other fire control units were the Information Coordination Central (ICC) and Platoon Command Post (PCP). The exterior appearance of the vans was nearly identical. In the foreground is an ICC and behind it a PCP of the 3d LAAM Bn on Onslow Beach, Camp Lejeune in 1986. In the background are a High Power Illuminator Radar and launcher. (NARA)

Product Improvement Program – the 1980s

After fielding IHAWK, the Army initiated a multi-phase Product Improvement Program (PIP) to incrementally modernize the system. Three upgrades, referred to as Phase or PIP I, II, and III, were completed. Additional upgrades were planned but not implemented before HAWK was retired from active service with the US military.

Accomplished from 1979 and 1981, PIP I increased IHAWK's low-level target acquisition capability by introducing a more powerful

and longer-range AN/MPQ-55 CWAR, enhancing the PAR's ability to see targets in ground clutter, and increasing the HIPIR's range to 100km (62 miles). New models of the ICC and PCP had redesigned computers and software that significantly speeded up the system's engagement response time, improved the sharing of target data with the battalion fire direction center and other HAWK firing units, and permitted interoperability with the soon-to-be fielded Patriot missile system. PIP II upgrades were fielded between 1983 and 1986. Modifications to the system included new radar and missile electronics to enhance HAWK's ECCM capabilities and a new AN/MPQ-57 HIPIR equipped with an optical tracking system to reduce the threat presented by Soviet anti-radiation missiles. Mounted between the radar's two antennae, the Tracking Adjunct System allowed visual target identification and tracking in daylight and clear weather without radiating the HIPIR and alerting an aircraft that it was being tracked. PIP II included two new faster missiles, the MIM-23C and MIM-23D models, which improved its ability to engage faster flying aircraft and increased the system's rate of fire by shortening flight time from launch to target.

The last upgrade, PIP III, was fielded from 1988–91. This major redesign streamlined the system's configuration and significantly increased its mobility and firepower. The BCC and ICC were replaced by the AN/MSW-21 Battery Command Post (BCP), and the PCP was upgraded to the AN/MSW-20. Both the BCP and new PCP were housed in the same type of shelter as the old PCP, but had new high-speed computers and tactical engagement consoles. The difference was that the PCP had only one tactical engagement console, while the BCP had two consoles that allowed the crew to control two firing sections much like the old BCC. Both the BCP and PCP could interface with the PAR for long-range surveillance and had computer software to enable the HAWK system to engage tactical ballistic missiles using target data from the Patriot missile system. A new model CWAR (the AN/MPQ-62) and new HIPIR (the AN/

MPQ-61) had enhanced digital signal processors to reduce ground clutter and counter the Soviets' latest electronic countermeasures. The AN/MPQ-50 PAR was unchanged and the ROR was deleted from the system. Lighter, high-speed data field wire replaced many of the system's cumbersome data cables. To increase firepower, the HIPIR was equipped with the Low-Altitude Simultaneous HAWK Engagement (LASHE) system. LASHE was designed to counter saturation attacks by allowing each HIPIR to engage up to six different targets simultaneously. New missiles, the MIM-23E and MIM-23F, with enhanced low-level and ECCM capabilities and a longer 46km (28 mile) range, replaced earlier models.

The PIP III system had several possible system configurations. Using the BCP allowed a full-battery configuration like previous versions of HAWK, with the BCP, PAR, CWAR, two HIPIRs, and launchers operating as one weapon system; or the flexibility to divide the system into two independent assault firing platoons, each equipped with a BCP or PCP, CWAR, HIPIR, and launchers. Additionally, the PAR could be used with the PCP to provide the assault firing platoon with its own long-range surveillance. The Army equipped its PIP III systems with PCPs, while the Marine Corps opted for BCPs.

Several other modernization programs were initiated in the mid-1980s as possible PIP upgrades. The Sparrow HAWK program sought to improve HAWK's firepower and mobility by modifying a launcher to fire either nine Sea Sparrow missiles or three HAWK missiles. The launcher was designed for use with LASHE and could carry several HAWK or Sparrow missiles while moving. However, the Sparrow missile's 19km (12 mile) range was insufficient against medium altitude targets, and the program was cancelled. The Agile Continuous Wave Acquisition Radar (ACWAR) program developed a trailer- or vehicle-mounted 3-Dimensional radar to replace all of HAWK's radars and provide advanced capabilities against electronic jamming, anti-radiation missiles, and tactical ballistic missiles; however, the radar's 20km (12½ mile) range was far too short. A classified program, sometimes called "Silent HAWK" or the "Alpha System," developed a passive-engagement system for PIP II and III systems for detecting and engaging Soviet aircraft carrying nuclear weapons. The system consisted of a special missile that could home in on emissions given off by an aircraft's own avionics, a sensor

B **FIRE CONTROL**

The Battery Control Central (BCC) controlled and directed the operations of the HAWK system. The BCC was operated by a crew of five: a tactical control officer (typically a lieutenant) and a tactical control assistant (either a sergeant or staff sergeant) who directed operations of the crew and selected targets for engagement, an azimuth speed operator who detected and reported low-flying targets acquired by the Continuous Wave Acquisition Radar (CWAR), and two fire control operators – one for each tracking radar – who tracked and engaged targets. During 20-minute "Hot Status" or other times of alert, the BCC was continually manned by at least one crew member. When at "Blazing Skies" or "Battle Stations," the BCC was fully manned. The Platoon Command Post served the same function as the BCC for the assault firing platoon.

The radarscope used by the tactical control officer and tactical control assistant displayed target information from the HAWK system's radars (see illustration). Information provided by the radars was supplemented by computer symbology generated by the Information Coordination Central or received via digital data link from higher headquarters fire distribution systems. The radarscope's display was updated by an electronic sweep that rotated around the scope every three seconds in synchronization with the rotation of the acquisition radars' antennas. Mounted above the radarscope was an electronic board that displayed information about the status of the HAWK system's components and missiles.

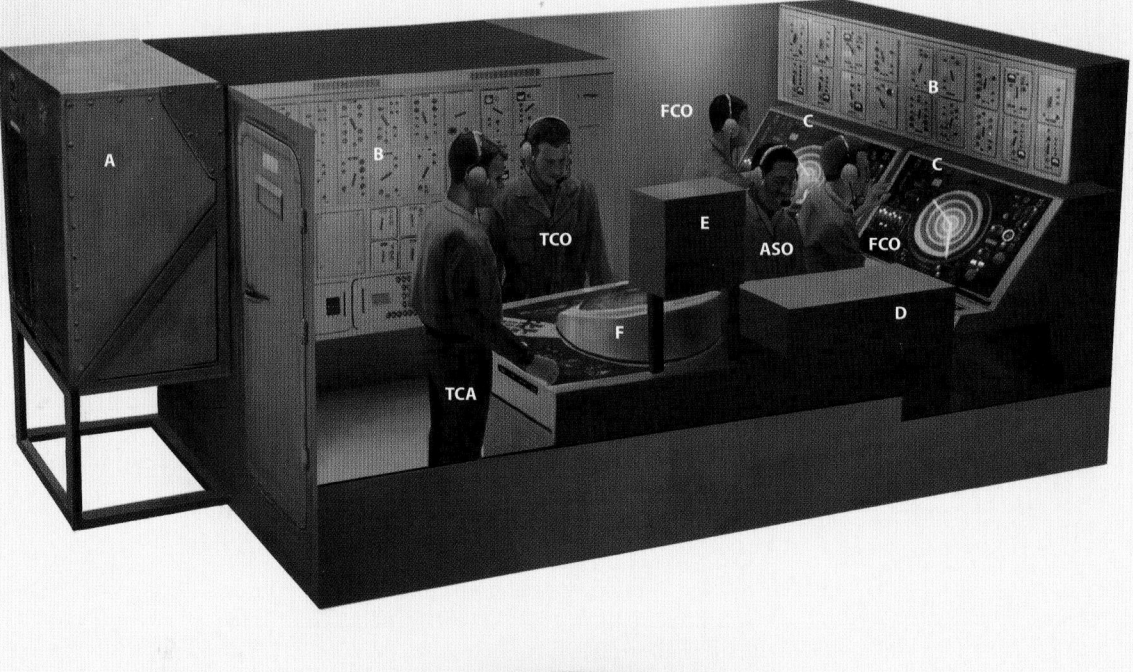

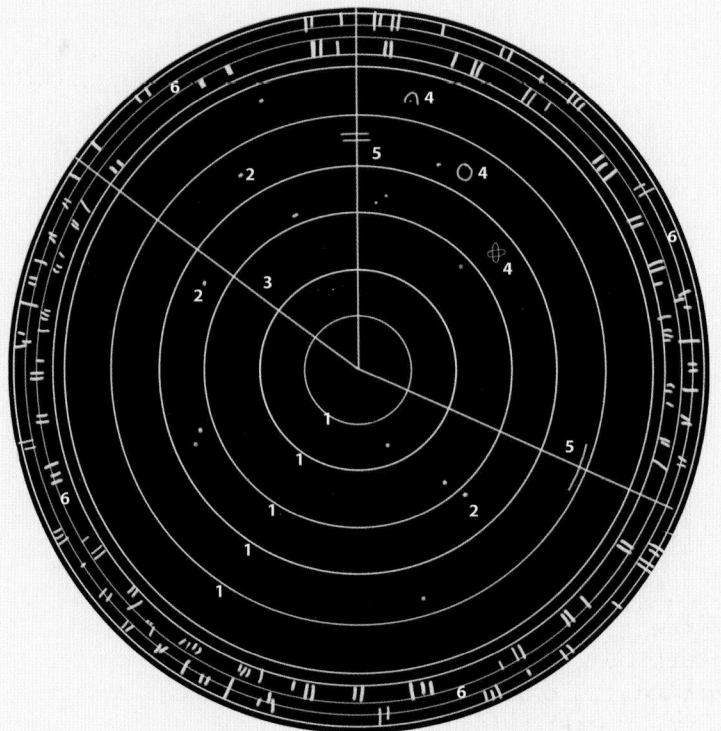

KEY

BCC components:

A. Air conditioner
B. Electronics racks
C. Firing consoles
D. CW target detection console
E. System status panel
F. Tactical control console

Crew members:

TCA. Tactical control assistant
TCO. Tactical control officer
FCO. Fire control operator
ASO. Azimuth speed operator

Radar scope display:

1. 20km range rings
2. PAR targets
3. Radar sweep
4. Computer generated target symbols
5. HIPIR azimuth and range markers
6. CWAR target display

The XM727 self-propelled launcher was a short-lived concept. On paper and during testing, the launcher seemed like a good way to increase HAWK's mobility, but once fielded with units in Germany, it was discovered that the launchers were prone to mechanical failure and sometimes even caught fire. (NARA)

disguised as a 60kw generator, a control unit mounted in the PCP, and a launcher modified to fire the special missile. The Alpha system was fielded by the US Army and deployed to the Gulf War.

Post-Cold War – the 1990s

After Soviet forces withdrew from Western Europe, US Army air defense artillery shifted focus from aircraft to ballistic missiles, and all active-duty HAWK battalions were inactivated, leaving HAWK in service with only the National Guard. Planned upgrades to the HAWK system had included faster and deadlier missiles, a new electro-optical tracker, and improved capabilities against tactical ballistic missiles, but when production of new equipment and missiles ended in 1993, PIP IV and possible PIP V upgrades were cancelled. Meanwhile, the Marine Corps continued to modify its PIP III systems under the HAWK Mobility Survivability Enhancement Program. This program changed the launcher so it could move while carrying missiles, installed a north-finding system to speed emplacement, and replaced the heavy data cables with data wire. These changes reduced the number of vehicles needed by a fire unit from 14 to seven and allowed the launchers to emplace up to 2km (1¼ miles) from the HIPIRs. The upgrades were implemented in 1995 and 1996, but then one year later the Marine Corps inactivated its last HAWK battalion.

HAWK units conducted annual live-fire missile exercises to maintain tactical proficiency. In Korea, missile live-fires were conducted by the 38th ADA Bde at the Special Eighth Army (SEA) Range located near Daechon Beach on the Yellow Sea. In the background are two Nike Hercules launchers. (NARA)

In addition to the United States, several other nations produced HAWK equipment. In the 1960s, Basic HAWK was co-manufactured in Europe by Société Européenne de Téléguidage, and from 1976–95, the NATO HAWK Production and Logistics Organization converted Basic HAWK to IHAWK and applied PIP upgrades. Japan's Mitsubishi Heavy Industries manufactured Basic and Improved HAWK equipment from

1968–78. In the late 1980s, Norway built its own version of the HAWK system called Norwegian Adapted HAWK (NOAH). Manufactured by Raytheon and Kongsberg Defence and Aerospace, NOAH replaced HAWK's acquisition radars and fire control units with a single trailer-mounted surveillance radar and a truck-mounted control unit. The NOAH system was in service from 1988–98 until replaced by the National or Norwegian Advanced Surface to Air Missile System (NASAMS) air defense system, which was a further development of NOAH. In the early 1990s, Germany, Netherlands, and Israel added long-range electro-optical systems to their HIPIRs. Denmark modified its HAWK systems by replacing the acquisition radars with a 3-Dimensional radar, replacing the BCC, ICC, and PCP with a new fire control center, and adding its own electro-optical tracking system to the HIPIR. Called the Danish Enhanced HAWK (DEHAWK), the system became operational in 2003, but was retired only two years later. In the mid-1990s, Sweden developed the Robotsystem 97 as an updated version of its HAWK systems.

The HAWK system continued to evolve in the late 1990s when Raytheon and Kongsberg jointly developed the HAWK Advanced Medium-Range Air-to-Air Missile (HAWK-AMRAAM) system, which combined PIP III equipment with NOAH's fire control center. This missile system further evolved into HAWK XXI. Several system configurations were possible, but a typical fire unit had an AN/MPQ-64 Sentinel 3-Dimensional radar for medium altitude and tactical ballistic missile detection and tracking, an AN/MPQ-62 CWAR for low altitude target detection, an updated AN/MPQ-61 HIPIR equipped with a passive infrared target system, a vehicle- or trailer-mounted Fire Direction Center with IFF, and as many as six missile launchers. The launchers could fire both HAWK MIM-23K or AIM-120 AMRAAM missiles. Three countries fielded HAWK XXI: Turkey bought eight firing units in 2005, Morocco purchased six in 2010, and Romania upgraded its eight PIP III systems, which it had purchased from the Netherlands in 2004, to HAWK XXI in 2018.

SYSTEM OPERATION

Unit organization
The US Army had two HAWK battalion organizational structures: square and triad. Square battalions, also adopted by the Marine Corps, consisted of four firing batteries, while triad battalions, which were originally formed with fielding of self-propelled HAWKs, had three batteries. When the Army retired the self-propelled HAWK in the mid-1970s, the former self-propelled HAWK battalions retained their triad organization. With the fielding of PIP III, Army HAWK square battalions were reduced to three firing batteries, while the Marine Corps reduced the number of batteries per battalion to two.

In addition to its firing batteries, each HAWK battalion had a battalion operations center (BOC) to operate its fire direction system, a communications platoon, a direct support unit for higher-echelon HAWK maintenance, and a specialized staff section to assist and evaluate the readiness of the firing batteries. In the 1980s, the direct support units were removed from many battalions and organized into separate ordnance companies. Army

Basic HAWK battalions had about 775 soldiers and 200 vehicles, while IHAWK battalions were larger with 875 soldiers and 230 vehicles. A PIP III battalion was significantly smaller, with about 500 soldiers and 140 vehicles. Marine Corps battalions were leaner than the Army's; for example, its Basic HAWK battalions had about 600 personnel.

In Basic HAWK batteries, the firing platoon operated the HAWK system. The AFU could be detached temporarily from the firing platoon as missions required. Firing batteries in square IHAWK battalions had two firing platoons: a base platoon equipped with a BCC, ICC, PAR, ROR, a tracking radar, and three launchers; and an assault fire platoon (formerly called the AFU) with a PCP, CWAR, a tracking radar, and three launchers. Each battery in a triad battalion had three firing platoons: one base (firing) platoon and two assault fire platoons. Although a triad battalion had fewer firing batteries, it had nine firing platoons versus a square battalion's eight. PIP III firing batteries had two assault fire platoons, each equipped with a PCP (or BCP for the Marine Corps), CWAR, a HIPIR, and three launchers. Army batteries assigned the PAR to one of the firing platoons, while each Marine Corps firing platoon had its own PAR.

In addition to its firing platoons, batteries had maintenance, communications, administrative, security, supply, and mess personnel. A typical HAWK square firing battery had approximately 150 soldiers and 35 vehicles, while a triad battery had about 175 soldiers and 45 vehicles. A PIP III firing battery was considerably smaller because it required fewer personnel to operate and maintain it.

Basic HAWK firing batteries were issued a trailer-mounted Guided Missile Test Shop to assemble, checkout, and maintain the unit's missiles. These soldiers are performing missile checks on the tactical site of C Btry, 6-71 ADA at Cam Ranh Bay, Vietnam in 1966. (NARA)

Operations

Deployed HAWK battalions maintained a constant state of alert with 24-hour manning. Firing batteries were assigned one of four alert states that specified the battery's readiness in terms of time required to bring the HAWK

C SELF-PROPELLED HAWK – GERMANY

Self-propelled HAWK was an attempt to improve Basic HAWK's mobility so it could operate closer – 10–15km (6–9 miles) – to front lines. The primary component of the self-propelled system was the XM727 tracked launcher, which was a modified tracked M548 cargo carrier with a XM78E3 missile launcher mounted in its cargo bed. The carrier's cab could carry the driver and three crewmen. During missile launch, a blast shield protected the rear of the cab, the carrier's engine, and onboard 60kw diesel generator (located directly behind the cab). An aluminum blast cover, stowed on top of the cab during movement, protected the windshield. Power and data cables were stored on an electric-powered cable reel assembly mounted on the rear of the carrier. For missile loading and firing operations, the carrier's suspension was locked in place to provide a stable platform. In this illustration, three soldiers are emplacing a self-propelled launcher.

A self-propelled HAWK assault fire unit (AFU) had fewer than half the vehicles of an AFU equipped with wheeled vehicles and launchers. Three self-propelled launchers, each carrying three missiles, could tow a High Power Illuminator Radar (1), a Platoon Command Post (2), and a Continuous Wave Acquisition Radar (3), which were all the components needed by the AFU to detect and engage enemy aircraft.

HAWK missiles were shipped to and from firing units in metal containers. Here, in 1960, crewmen are using a 2½-ton crane truck to move a container in the missile storage area of A Btry, 3-7 Arty near Schweinfurt, Germany. (NARA)

system into operation: 20-Minute, 3-Hour, 12-Hour, or 12-Hour Released. In the 1960s, instead of 20-Minute status, a 5-Minute standby status was used. Alert states were rotated weekly within a battalion, with one firing battery holding each state. The highest state of alert was 20-Minute or "Hot Status." The 20-Minute battery kept its HAWK system fully operational and constantly manned. All equipment was powered, surveillance and tracking radars were energized, key codes were loaded into the IFF, and a tactical data link to the battalion's fire direction system was maintained. When alerted, the crew brought the HAWK system to either "Blazing Skies" or "Battle Stations" ready-to-fire status. At Blazing Skies, the term used for peacetime training and evaluations, the system was fully operational, but the missiles were not armed or ready to fire. For Battle Stations, missiles were armed and could be fired with the push of a button by the BCC crew. The 3-Hour battery was also operational and manned as a back-up to the 20-Minute battery, whereas the HAWK systems of the 12-Hour batteries were partially operational and sometimes shut down so unit personnel could conduct comprehensive system maintenance or unit training. During times of increased tension, 12-Hour batteries were ordered to a higher state of alert.

To maintain readiness, HAWK units underwent frequent evaluations. Most common was the operational readiness evaluation, conducted to maintain crew combat readiness. During the evaluation, the crew of a Hot Status

HAWK radars required frequent periodic checks and adjustments to ensure proper performance. This radar mechanic has removed a cover on the receiver antenna of an AN/MPQ-46 HIPIR to conduct electronic checks during an exercise at Fort Bliss, Texas in 1982. (NARA)

battery had to prove it could bring its HAWK system to an operational, ready-to-fire status in 20 minutes or less. To train air battle procedures, weekly or monthly air defense exercises were conducted with higher headquarters' fire direction centers. Each year, most batteries live-fired a missile during its annual service practice and, every two years, battalions underwent a week-long tactical evaluation to assess its ability to operate and sustain itself in a simulated combat environment to include ground defense; nuclear, biological, and chemical defense; and tactical mobility.

To engage targets successfully, the HAWK system's radars and launchers had to be aligned with each other to correlate target location and data. This Marine in A Btry, 3d LAAM Bn is using an alignment telescope to align a Range-Only Radar during an exercise at Marine Corps Air Station Cherry Point in 1979. (NARA)

The HAWK system required continual maintenance conducted on a daily, weekly, monthly, and quarterly schedule, as prescribed by technical manuals. When a malfunction occurred, maintenance personnel diagnosed the problem and replaced faulty components from the unit's stock of repair parts. When in Hot Status, unit personnel were required to quickly repair critical system malfunctions or report the unit as non-operational and drop out of Hot Status until the malfunction was resolved. For lengthier or more difficult repairs, the battalion's direct support maintenance unit dispatched contact teams to assist the firing batteries. The direct support unit maintained additional spare parts on hand, special tools and test equipment, and if needed, spare (also known as float) radars and equipment to temporarily replace non-operational pieces at the firing batteries. HAWK equipment needing extensive repair was sent to a general support ordnance company.

HAWK battalions operated as part of an air defense network. Tactical command and control was accomplished by air defense group (brigade) and battalion fire direction centers that monitored the air picture, provided early warning to fire units, and, when necessary, assigned targets to fire units. Fire direction centers transmitted information between each other and the firing units by secure ultra-high frequency multi-channel radio digital data links. Basic HAWK battalion fire direction centers used the mobile AN/TSQ-38, Battalion Operations Central. Mounted on a M35 2½-ton cargo truck, this large metal equipment shelter housed two consoles with radarscopes for displaying an air picture from its own radar, subordinate HAWK fire units, and other fire direction centers in the air defense network. Target surveillance was provided by either an AN/GSS-1, AN/GSS-7, or AN/TPS-1D search radar, which could detect targets up to 297km (186 miles). The radar and IFF equipment were also mounted on a M35 2½-ton cargo truck. In the late 1970s, the AN/TSQ-38 was replaced by the AN/TSQ-73 Missile Minder. This solid-state electronics version of the AN/TSQ-38 allowed firing batteries to engage targets detected by other fire direction centers and HAWK batteries. The AN/TSQ-73 and associated equipment were mounted on M814 5-ton cargo trucks. Battalion AN/TSQ-73s used the AN/GSS-1 radar until replaced by the HAWK PAR in the late 1980s.

Mobility was a key capability of the HAWK system. The system was designed to be march-ordered in 30 minutes or less and then emplaced and

HAWK batteries in Germany often underwent mobility exercises. These trucks of B Btry, 6-62 ADA are crossing a pontoon bridge during a tactical evaluation in 1971. The two missile trucks are towing missile pallet trailers. (NARA)

ready to fire within 45 minutes. When changing locations, firing batteries moved one firing platoon at a time so that part of the battery always provided air defense coverage. In Germany, triad batteries had additional flexibility because of the third firing platoon. Before a firing unit moved into a new location, a reconnaissance, selection, and occupation of position (RSOP) team selected and prepared the new site. This team consisted of a dozen or so personnel with equipment needed to secure the new position and site each piece of HAWK equipment before the battery's arrival. The ideal field position was somewhat level, located on high ground, and had cover and concealment. An area of about 200 by 400 meters (20 acres) was needed for an entire HAWK battery, and a 100 by 200-meter (5-acre) area was needed for an assault firing platoon. Additional space was needed for unit vehicles and support equipment.

OPERATIONAL HISTORY

From 1959–62, the US Army formed 21 HAWK battalions and two separate batteries. The first battalion (5-57 Arty) served as an instructional battalion at Fort Bliss. Fifteen battalions were deployed overseas: four to Korea, nine to Germany, and two to Okinawa. The two separate firing batteries were sent to Panama. The remaining battalions were stationed in the United States: three at Fort Bliss as contingency forces and one each at Fort Lewis, Washington, and Fort Meade, Maryland to augment Nike Hercules missile battalions providing air defense of the Continental United States. Army HAWK battalion lineage and designations were complex. The first battalions, formed under the Artillery Branch were designated as missile artillery battalions; for example, 6th Missile Battalion, 52d Artillery (abbreviated as 6th Msl Bn, 52d Arty or 6-52 Arty). To differentiate HAWK battalions from other types of missile battalions, "HAWK" was often added to the unit title; for example, 6th Msl Bn (HAWK), 52d Arty. Soon after Air Defense Artillery (ADA) became its own branch of the Army in 1968, battalion titles changed to "Air Defense Artillery," such as 6th Battalion, 52d Air Defense Artillery (abbreviated as 6th Bn, 52d ADA, or just 6-52 ADA), and then in the 1970s and again in the 1980s, several HAWK battalions were redesignated (reflagged) as part of a realignment of ADA regimental affiliations.

Two HAWK batteries were deployed to the Panama Canal Zone in 1960 and assigned to 4-517 Arty. This tower-mounted Basic HAWK Pulse Acquisition Radar of D Battery was located at Fort Sherman in the Canal Zone. (NARA)

US Army HAWK Battalions		
Battalion	**Period of Activation**	**Primary Station(s)**
1-1 ADA	1972 (Reflagged from 6-562) – 1986 (Reflagged as 3-52) 1986 (Reflagged from 2-62) – 1991	Germany
2-1 ADA	1987 (Reflagged from 1-7) – 1994	Ft Bliss, Persian Gulf
3-1 ADA	1987 (Reflagged from 1-65) – 1993	Ft Bliss, Ft Hood
4-1 ADA	1987 (Reflagged from 3-59) – 1993	Germany
8-1 Arty	1960–70	Okinawa
1-2 ADA	1972 (Reflagged from 7-2) – 1981	Korea
2-2 ADA	1972 (Reflagged from 6-517) – 1985	Germany
7-2 Arty	1960–72 (Reflagged as 1-2)	Korea
8-3 Arty	1961–73	Okinawa
1-4 ADA	1978–88 (Reflagged as 1-52)	Ft Lewis
7-5 Arty	1962–71	Korea
1-7 ADA	1972 (Reflagged from 8-7) – 1987 (Reflagged as 2-1)	Ft Bliss
3-7 Arty	1960–86	Germany
8-7 Arty	1962–72 (Reflagged as 1-7)	Ft Meade, Ft Bliss
8-15 Arty	1961–71 (Reflagged as 3-68)	Ft Lewis, Homestead
1-44 ADA	1972 (Reflagged from 6-44) – 1980	Korea
6-44 Arty	1962–72 (Reflagged as 1-44)	Korea
2-51 ADA	1978–84	Ft Riley
1-52 ADA	1988 (Reflagged from 1-4) – 1993	Ft Lewis
2-52 ADA	1988 (Reflagged from 3-68) – 1993	Ft Bragg, Persian Gulf
3-52 ADA	1986 (Reflagged from 1-1) – 1992	Germany
6-52 Arty	1960–93	Germany, Persian Gulf
2-55 ADA	1972 (Reflagged from 6-61) – 1985	Ft Bliss
6-56 Arty	1962–69	Ft Bliss, Vietnam
2-57 ADA	1972 (Reflagged from 4-57) – 1985	Germany
4-57 Arty	1960–72 (Reflagged as 2-57)	Germany
5-57 Arty	1959–82	Ft Bliss
3-59 ADA	1972 (Reflagged from 6-59) – 1987 (Reflagged as 4-1)	Germany
6-59 Arty	1960–72 (Reflagged as 3-59)	Germany
3-60 ADA	1972 (Reflagged from 6-60) – 1990	Germany
6-60 Arty	1961–72 (Reflagged as 3-60)	Germany
6-61 Arty	1961–72 (Reflagged as 2-55)	Germany, Ft Bliss
2-62 ADA	1972 (Reflagged from 6-62) – 1986 (Reflagged as 1-1)	Germany
6-62 Arty	1961–72 (Reflagged as 2-62)	Germany
1-65 ADA	1972 (Reflagged from 6-65) – 1987 (Reflagged as 3-1)	Key West, Ft Bliss
6-65 Arty	1961–72 (Reflagged as 1-65)	Ft Meade, Key West
3-68 ADA	1971 (Reflagged from 8-15) – 1988 (Reflagged as 2-52)	Homestead, Ft Bragg
2-71 Arty	1960–82	Korea
6-71 Arty	1962–70	Ft Bliss, Vietnam
2-174 ADA	1990–97	McConnelsville, OH
7-200 ADA	1986–96	Rio Rancho, NM
1-263 ADA	1991–95	Anderson, SC
2-265 ADA	1990–95	Titusville, FL
4-517 Arty	1960–70	Panama
6-517 Arty	1962–72 (Reflagged as 2-2)	Germany
6-562 Arty	1962–72 (Reflagged as 1-1)	Germany
Note: US Army battalions are listed as Arty or ADA based upon their designation at time of activation		

In 1960 and 1961, the Marine Corps activated the 1st, 2d, and 3d Light Antiaircraft Missile Battalions (LAAM Bns) at Marine Corps Base Twentynine Palms, California, and a reserve 4th LAAM Bn in Fresno, California. After 1st and 2d LAAM Bns deployed to Vietnam, 5th LAAM Bn was formed at Marine Corps Air Station Yuma, Arizona, to train Marines being assigned to a HAWK battalion in Vietnam, but was inactivated in 1969 after 1st and 2d LAAM Bns returned to the United States.

US Marine Corps HAWK Battalions		
Battalion	**Period of Activation**	**Primary Station(s)**
1st LAAM Bn	1960–70	29 Palms, Vietnam
	1987–90	Okinawa
	1994 (Reflagged from 2d LAAM Bn) – 1997	Yuma
2d LAAM Bn	1960–94 (Reflagged as 1st LAAM)	29 Palms, Vietnam, Yuma, Persian Gulf
3d LAAM Bn	1961–94	29 Palms, Cherry Point, Cuba, Persian Gulf
4th LAAM Bn	1961–97	Fresno, CA
5th LAAM Bn	1966–69	Yuma

Panama

To protect the strategically important Panama Canal from attack by Soviet aircraft based in Cuba, two HAWK firing batteries were deployed to the Canal Zone in 1960. Assigned to 4th Bn (HAWK-AW), 517th Arty, these batteries were part of a unique composite missile and automatic weapons battalion, the only battalion of its type in the US Army. A and B Batteries (Btrys) were equipped with 40mm self-propelled M42 "Duster" antiaircraft guns, while C and D Btrys were HAWK. The HAWK batteries

These soldiers are conducting a launcher function check using a launcher test set (the box on the ground to the right of the launcher). The missiles have been removed and the launcher's boom is at maximum elevation. The photo was taken in 1963 on the tactical site of A Btry, 6-44 Arty in Korea. (NARA)

were emplaced at opposite ends of the Canal Zone on permanent tactical sites: D Btry located on the Atlantic side on a hill above Fort Sherman, and C Btry on the Pacific side at Fort Grant on Flamenco Island in the Bay of Panama. Although the radars were mounted on towers, the batteries had tactical vehicles to move if required. D Btry became operational in October 1960 – the first HAWK battery in the world to assume an operational 24-hour alert status. However, HAWK's time in Panama was short-lived. In 1968, C and D Btrys were inactivated, followed by the remainder of the battalion in 1970.

Korea

To bolster defenses against North Korean aggression, the US Army deployed four HAWK battalions to the Republic of Korea (ROK) in the early 1960s to defend the Demilitarized Zone (DMZ) and protect US military bases from potential North Korean or Chinese air attack. The first battalion (2-71 Arty) arrived in 1960 and occupied positions around Uijongbu near the DMZ. Three more HAWK battalions soon followed: 7-2 Arty deployed to the Shihung area south of Seoul in 1961, 7-5 Arty moved into the Chuncheon area to defend the eastern part of the DMZ in 1962, and 6-44 Arty deployed to the vicinity of Kwang Chun in 1963 to protect the west coast against air attack across the Yellow Sea. All four HAWK battalions, along with a Nike Hercules battalion, were placed under the command of the 38th Artillery Brigade.

Most HAWK tactical sites in Korea were located on remote mountain tops. This site, photographed in 1967, belonged to D Btry, 2-71 Arty and was located only 23km (14 miles) from the demilitarized zone between North and South Korea. (George Blanchette)

TACTICAL SITE, GERMANY

A Battery, 6th Battalion, 52nd Air Defense Artillery, Würzburg, Germany

HAWK tactical sites were designed and built to accommodate the operational requirements of the HAWK missile system. Each site covered about 30 acres and contained the facilities unit personnel needed to operate and maintain the system. The shape and size of the site was determined by terrain. The radars were emplaced on earthen berms (or sometimes towers), to improve coverage and reduce ground clutter. Launchers were emplaced facing the expected direction of enemy air attack (called the primary target line), and often separated by earthen berms to shield adjacent launchers from missile back blast. Missiles not loaded on launchers were kept in a storage area surrounded by protective berms. The fire control units were centrally located between the radars and launchers. The site's buildings included a ready building with barracks, mess hall, arms room, and an office; a maintenance building for repairing and servicing HAWK equipment and components; a motor pool for servicing vehicles; a communications building; and sheds for the generators. A chain link fence surrounded the site.

The tactical site depicted in this illustration was occupied by A Battery, 6th Battalion, 52nd Air Defense Artillery near Würzburg, Germany, from 1962–92. When first occupied by the battery, unit personnel named the site after Specialist 5 Steve Aston, who had been recently killed in a vehicle accident with an Army 2½-ton truck.

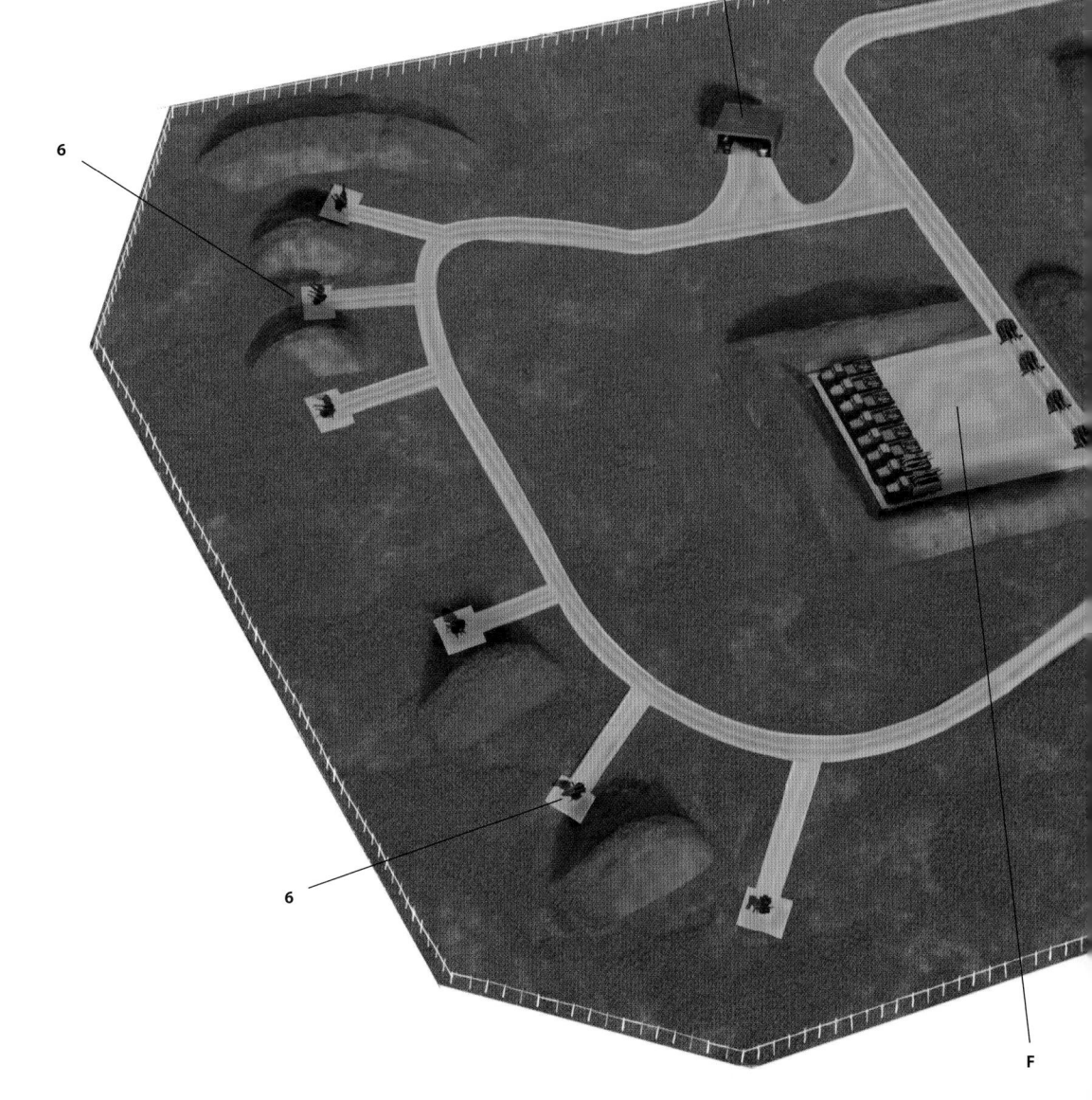

The tactical site

A. Entrance
B. Motor pool
C. Crew building
D. HAWK maintenance building
E. Generator shed
F. Missile storage area

The HAWK system

1. Pulse Acquisition Radar
2. Continuous Wave Acquisition Radar
3. Range-Only Radar
4. High Power Illuminator Radar
5. Control vans (ICC, BCC, & PCP)
6. Missile launchers

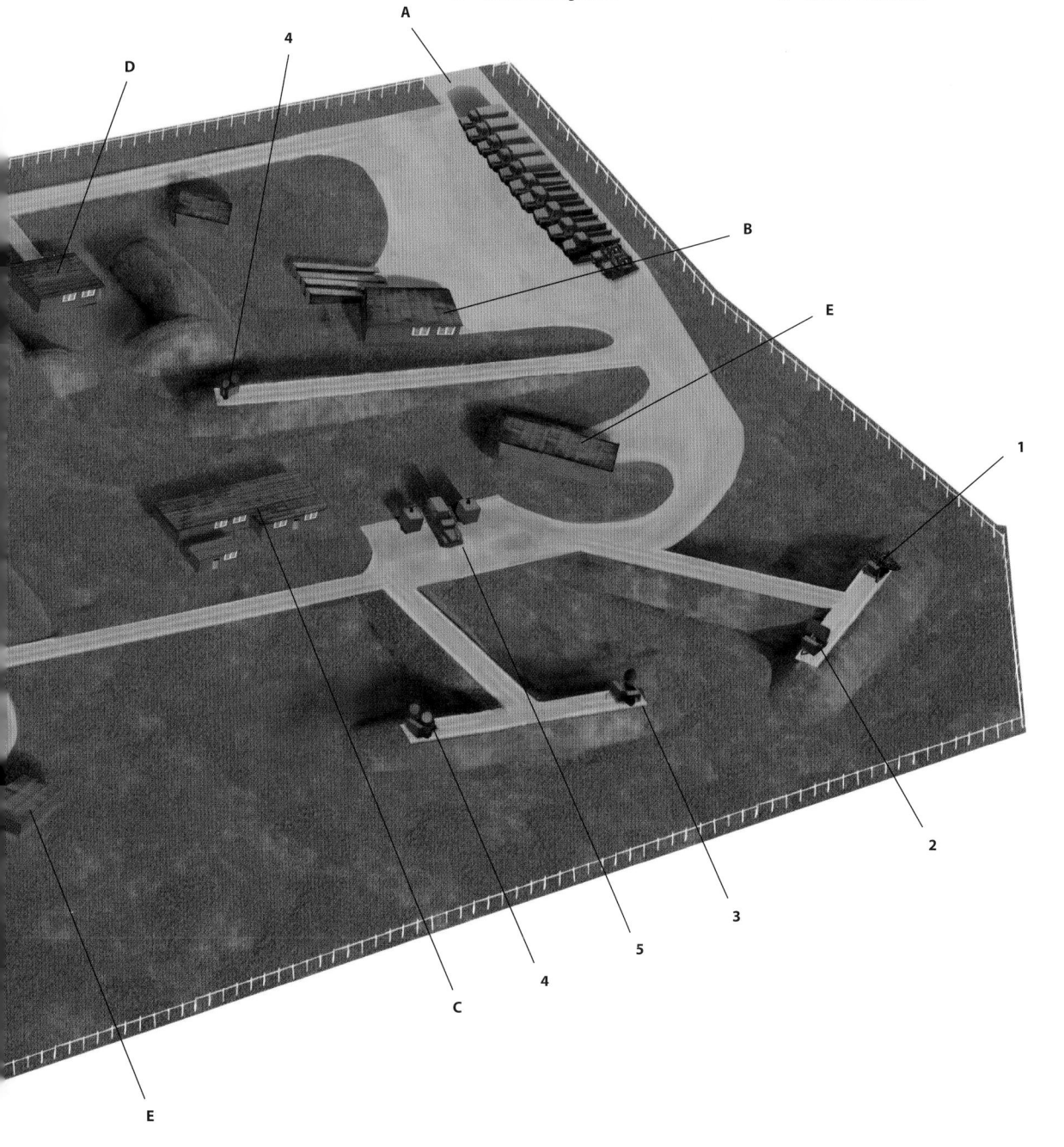

Two US Army HAWK battalions were stationed on Okinawa in the 1960s. Here, a crewman of C Btry, 8-1 ADA operates a launcher's control unit to conduct an azimuth and elevation test during an alert. (M. Romanych)

Korea posed an exceptionally challenging operating environment. HAWK tactical sites were widely dispersed and located on remote mountain-top positions, with B Btry, 7-5 Arty, sited at 4,600ft (1,400m) elevation on Mae-Bong Mountain, making it the highest HAWK tactical site in the world. Created by bulldozing mountain tops, these sites were accessible only by narrow, winding dirt roads. To overcome the sites' remoteness, 38th Artillery Brigade often used utility aircraft and helicopters to transport essential personnel and repair parts. Even when the sites were improved with permanent structures, living conditions remained austere, with no running water and electricity provided by tactical generators.

North and South Korea were still technically at war and waging unconventional warfare against each other. The tense military situation necessitated the 38th Artillery Brigade maintain a wartime readiness posture to counter North Korean espionage, the threat of commando attack, and for HAWK sites near the DMZ, the risk of an artillery attack or air strike. In response, the brigade hardened its tactical sites. Double fence perimeters with lighting were installed, vegetation defoliated, defensive fighting positions improved, and revetments constructed to protect personnel and equipment from small-arms fire, artillery fire, or air strike. To deter infiltration, many tactical sites were guarded by sentry dogs.

In addition to the ground threat, the North Korean Air Force and Chinese Air Force aggressively probed US air defenses with aircraft overflights, perhaps trying to provoke an international incident. Because of these incursions, HAWK batteries endured prolonged periods of high alert status, frequently went to battle stations, and on several occasions came close to launching a missile on North Korean or Chinese aircraft. Most affected

One of A Btry, 6-71 Arty's AN/MPQ-39 HIPIRs overlooking Cam Ranh Bay, Vietnam in 1965. When not actively tracking targets, tracking radar antennas were usually set on a predetermined azimuth and elevated to 45 degrees to reduce the radiation hazard to unit personnel. (NARA)

were 2-71 Arty's firing batteries, which were only a few seconds away from the DMZ by aircraft, and 6-44 Arty, which guarded the airspace along the Yellow Sea coast. Although the firing batteries were mobile, mountainous terrain constrained movement from tactical sites to field positions. Recognizing this problem, 38th Artillery Brigade designated 7-2 and 6-44 Arty as mobile units in 1968 and issued additional vehicles and communication equipment to enhance the battalions' ability to move. The brigade also conducted airmobile training, including the airlift of entire HAWK batteries by CH-47 helicopters. These operational improvements paid off in 1976 when North Korean soldiers killed two US Army officers in the DMZ, and D Btry, 1-2 ADA (formerly 7-2 Arty) quickly deployed to strengthen air defense coverage along the DMZ.

Basic HAWK missiles required more periodic checks than the later model missiles produced for IHAWK. These soldiers of B Btry, 6-71 Arty are using a multimeter to conduct a stray voltage check on Basic HAWK missiles at Cam Ranh Bay, Vietnam in 1966. (NARA)

Gradually, the ROK Army assumed responsibility for all HAWK operations. It formed its own HAWK battalion in 1965 and ten years later, during a US force reduction, took over 7-5 Arty's tactical sites and equipment when the battalion was inactivated. Then from 1977–82, during another US troop reduction, 38th ADA Brigade turned its remaining battalions and mission over to the ROK Army. The Nike Hercules battalion was transferred first, followed by 1-44 ADA (formerly 6-44 Arty) in 1980, and 1-2 ADA (formerly 7-2 Arty), and the remainder of 38th ADA Brigade in 1981. The last battalion (2-71 ADA) was transferred and inactivated in 1982, ending 23 years of US HAWK presence in Korea. However, the ROK military continued to operate four HAWK battalions until replaced by the Cheongung SAM system in 2021.

Okinawa and the Far East

In 1961, two HAWK battalions (8-1 and 8-3 Arty) deployed to Okinawa to reinforce two Nike Hercules battalions of the 30th Artillery Brigade that were protecting US military bases against Chinese air attack. The HAWK batteries were dispersed among the Nike batteries, with 8-1 Arty, headquartered at Machinato, occupying sites in the central part of the island and 8-3 Arty, headquartered in Naha, manning sites on the southern tip of Okinawa and nearby Tokashiki Island. None of the firing batteries were mobile. The tactical sites had permanent buildings, asphalt roads, and earthen berms that resembled the fixed air defense missile sites built in the US for continental air defense.

As the threat of Chinese air attack diminished, 8-1 Arty gradually inactivated between 1968 and 1970, leaving 8-3 Arty as the sole HAWK battalion on Okinawa. Two years later, when the Japanese government took over administration of the island, 30th Artillery Brigade transitioned its mission and units to the Japanese Ground Self-Defense Force, which already had several HAWK and Nike Hercules battalions operating elsewhere in Japan. By 1973, after 13 years on 24-hour alert status, 8-3 ADA and the

HAWK battalions possessed an organic long-range surveillance radar to provide early warning for the firing batteries. This AN/GSS-1 radar of 8-15 Arty, photographed in 1963 in Homestead, Florida, was mounted on top of the headquarters building. In the background is a "fly swatter" antenna of the multichannel radio system used to establish voice and data links between the battalion's fire direction center and its firing batteries. (NARA)

remainder of 30th ADA Brigade were inactivated. US HAWK missiles briefly returned to Okinawa in 1987 when 1st LAAM Bn was assigned to III Marine Expeditionary Force for three years. Japan's HAWK systems remained active until replaced in 2005 by the Type 03 Chu-SAM.

Two other Far East nations also received and operated HAWK systems: Taiwan acquired two battalions in the 1960s, later expanded to four, and in 1982 Singapore received four batteries. Both countries upgraded these systems and operated them for decades until 2020.

Florida and the Cuban Missile Crisis

After a U-2 reconnaissance flight photographed nuclear-armed Soviet ballistic missiles in Cuba on October 15, 1962, the US rapidly deployed troops and aircraft to Florida for a possible invasion of Cuba. To provide air defense coverage, two HAWK battalions (6-65 and 8-15 Arty) and a Nike Hercules battalion were deployed to defend airbases and unit staging areas from nuclear-capable Soviet IL-28 bombers and MiG-21 fighters, which could fly to Florida from Cuba in approximately five minutes. These air defense battalions were placed under command of the 13th Artillery Group. On short notice, 6-65th Arty deployed from Fort Meade to Key West, while 8-15 Arty deployed from Fort Lewis to protect Patrick, MacDill, and Homestead Air Force Bases in central Florida. Both battalions became operational on October 26, two days before the crisis formally ended. At the same time, the Marine Corps moved 3d LAAM Bn from Twentynine Palms to Marine Corps Air Station Cherry Point in North Carolina. From there, its C Btry was airlifted to Guantanamo Bay Naval Base in Cuba and emplaced on the base's highest point. The battery became operational on October 26 and provided air defense coverage for several weeks before redeploying to the US in mid-December.

Although the crisis eased on October 28 when the Soviets agreed to remove its nuclear weapons from Cuba, the 13th Artillery Group and its three battalions were kept on alert to deter and protect against air attack from Cuba. Five months later, 13th Artillery Group was permanently assigned to the defense of Florida and integrated into the air defense of the Continental US,

A launcher section of C Btry, 3d LAAM Bn at Guantanamo Bay Naval Base in Cuba. Airlifted into the base on October 26, 1962, the battery provided air defense protection for about 45 days during the Cuban Missile Crisis. (M. Romanych)

with 6-65 Arty staying on Key West while 8-15 Arty repositioned all its firing batteries around Homestead Air Force Base. For nearly three years, the firing batteries occupied temporary field positions until permanent tactical sites were completed in the summer of 1965. These permanent sites had hardstand facilities for crews and equipment, electric power provided by the civilian power grid, and towers to elevate the radars above nearby buildings and trees. After occupying the sites, both battalions became semi-mobile, with limited organic capability to move their BOCs and firing batteries.

Southern Florida's air defenses, including the HAWK battalions, were occasionally probed by Cuban MiGs. Surprisingly, on three occasions in the 1960s, Cuban aircraft – an Mi-4 helicopter in 1964, an AN-2 biplane in 1968, and a MiG-17 in 1969 – piloted by defectors or refugees flew from Cuba to Florida and landed without detection by either Air Force or Army surveillance radars. Each of these embarrassing incidents highlighted air defense vulnerabilities and prompted procedural changes to surveillance and control of southern Florida's airspace.

In 1968, 13th Artillery Group was redesignated as 47th Artillery Brigade, and three years later as 31st ADA Brigade. In the early 1970s, the battalions were reflagged, with 6-65 Arty and 8-15 Arty becoming 1-65 ADA and 3-68 ADA respectively. After 18 years, when the threat of air attack ended, Army air defense operations in Florida ceased and 1-65 ADA moved to Fort Bliss, 3-68 ADA moved to Fort Bragg, and the 31st ADA Brigade was inactivated.

Transferring missiles to and from the launchers was a challenging operation requiring training and skill. This missile transfer is taking place in 1975 on the tactical site of A Btry, 3-68 ADA near Homestead Air Force Base, Florida. In the background is a tracking radar tower. (M. Romanych)

Vietnam

When the Viet Cong attacked a US military compound at Pleiku on February 7, 1965, President Lyndon B. Johnson immediately ordered retaliatory air strikes and deployment of a HAWK battalion to South Vietnam. Because the 1st LAAM Bn was already positioned on Okinawa for possible deployment to

Vietnam, its A Btry was airlifted to Da Nang Airbase the day after the Pleiku attack. This HAWK battery was the first US combat unit to enter Vietnam, and upon arrival, quickly established air defense coverage of the airbase. A few weeks later, A Btry was joined by the rest of 1st LAAM Bn which arrived by sealift. In September, 2d LAAM Bn arrived from Twentynine Palms, deployed to Chu Lai Airbase, and became operational on October 2. Both HAWK battalions were placed under command of the 1st Marine Aircraft Wing.

Two Army HAWK battalions (6-56 and 6-71 Arty) arrived in October 1965 from Fort Bliss. To protect Tan Son Nhut and Bien Hoa Airbases, 6-56 Arty established tactical sites near Saigon, while 6-71 Arty initially went to Qui Nhon Airbase north of Saigon, but was then redeployed to protect air and naval installations in the Cam Ranh Bay and Na Trang areas. By mid-November, both battalions were operational and under the command of the 97th Artillery Group.

The first HAWK sites were field positions, but within a year, most were improved with buildings, radar berms, and launcher pads. Soon after, many HAWK batteries were targeted by Viet Cong sniper fire, ground attacks, and artillery fire, occasionally inflicting casualties and damage. When attacks increased in 1967, 1st Marine Aircraft Wing and 97th Artillery Group fortified their tactical sites to protect personnel and equipment. Fortification measures varied by battalion, but generally included revetments for HAWK equipment and buildings, personnel bunkers, perimeter barricades and lights, listening posts, Claymore mines, seismic intrusion detectors, and removal of perimeter vegetation.

Basic HAWK missile launchers on the tactical site of A Btry, 6-60 Arty near Grafenwöhr, Germany in 1964. HAWK battalions did not intend to fight from the tactical sites unless defending against a surprise attack and planned to move to predesignated field sites in the event of war. (NARA).

A self-propelled launcher towing a tracking radar during a REFORGER (Return of Forces to Germany) exercise in 1970. Only four battalions in Germany and one battalion at Fort Bliss were equipped with self-propelled launchers. (NARA)

During the 1968 Tet Offensive, the Viet Cong attacked several HAWK sites. Hardest hit was A Btry, 1st LAAM Bn, which had eight Marines wounded, a launcher with missiles destroyed, and another fourteen missiles, two data cables, and a J-Box damaged, and B Btry, 2d LAAM Bn, which was struck by a rocket attack that damaged a launcher, causing three missiles to ignite and launch, but inflicted no casualties. The next year, attacks against 1st LAAM Bn continued culminating in an attack against C Btry which killed one and wounded two Marines, destroyed two launchers and 12 missiles, and damaged two radars.

NATO's HAWK Belt in Germany had 96 battery tactical sites, eight of which had towers to improve radar coverage. This photograph of the German 1./Flugabwehrraketenbataillon 36 site near Ebersdorf shows all five of the battery's radars and its Identification Friend or Foe antenna mounted on towers. (Ulrich Kroemer)

Although the North Vietnamese never mounted an air attack, HAWK firing batteries remained vigilant, spending lengthy periods on 5-minute alert status whenever unidentified aircraft crossed the DMZ. To maintain combat readiness, the battalions conducted readiness tests and missile live-fires over the South China Sea. By 1968, the only serious air threat was North Vietnamese helicopters operating along the DMZ and Cambodian border. To counter this, 2d LAAM Bn tested the feasibility of using the HAWK system to detect low-flying helicopters and pass information needed for A-4 aircraft to intercept the helicopters. After the tests were successful, 2d LAAM Bn prepared to move its B Btry closer to the DMZ, but the operation was never implemented.

As part of a US troop reduction, 2d LAAM Bn returned to Twentynine Palms and 6-71 Arty to Fort Bliss in the autumn of 1968. At the same time, 97th Artillery Group was inactivated, and 6-56 Arty (minus its C Btry) was assigned to the 23rd Infantry Division (Americal Division) and deployed to Chu Lai to replace 2d LAAM Bn. C Btry, 6-56 Arty remained at Tan Son Nhut Airbase awaiting deployment to Thailand. However, that deployment was cancelled, and in March 1969, the battery deployed its radars and BCC to Landing Zone (LZ) Oasis near the Cambodian border for Operation *Western Pistol*, a special project to locate and identify helicopters operating in the region. During the operation, in a unique employment of the HAWK system, C Btry used its BCC and radars to direct 8in. howitzer fire against helicopter landing zones. In retaliation, the North Vietnamese Army attacked LZ Oasis, destroying C Btry's CWAR and damaging the PAR. Fortunately, the battery suffered no casualties. When Operation *Western Pistol* ended in June, C Btry was inactivated in country. In August, 1st LAAM Bn redeployed to Twentynine Palms and 6-56 Arty returned to Fort Bliss, and both inactivated soon afterwards.

During their four and a half years in Vietnam, HAWK battalions never fired a missile at Vietnamese aircraft. Nevertheless, HAWK posed a credible deterrent to North Vietnamese air attack, served an important role in the US Air Force air traffic surveillance network, and allowed Air Force and Marine fighter aircraft to focus on offensive air missions over North Vietnam instead of flying defensive patrols in the south.

Germany and Cold War Europe

In the late 1950s and early 1960s, as tensions grew in Europe between the Soviet Union and Western Powers, NATO planned to create a belt of HAWK missile units extending across Germany, from Denmark to Austria, to protect against air attack by the Warsaw Pact. This belt was divided into 26 battalion sectors assigned to either the United States, Germany, France, Belgium, or the Netherlands. The US part of the "HAWK Belt" was originally planned to stretch from Giessen to Regensburg, but was temporarily extended north towards Kassel and south to Munich until Germany and France could deploy their HAWK units into those sectors.

The US Army deployed nine HAWK battalions to Germany. The first two battalions (6-52 and 4-57 Arty) arrived in December 1960 to occupy positions near Würzburg and Ansbach. Three months later, 6-52 Arty became the first operational HAWK battalion in Germany. In 1961, another three battalions were stationed in the US part of the HAWK Belt (3-7 Arty at Schweinfurt, 6-59 Arty at Hanau, and 6-60 Arty at Grafenwöhr), and two battalions occupied French and German sectors near Munich (6-61 Arty at Landshut and 6-62 Arty at Dachau). The last US HAWK battalions arrived in 1962: 6-517 Arty moved to the Giessen area and 6-562 Arty went to Butzbach. With all nine US battalions operational, HAWK became the mainstay of NATO air defense for the next 25 years.

Other NATO nations began occupying their assigned sectors in 1964: Germany fielded nine battalions in northern and southern Germany, the Dutch deployed three battalions near Hanover, and two Belgian battalions took positions near Kassel. In 1965, the French moved two HAWK battalions to the Munich area, allowing the US to reposition 6-62 Arty to

E **MISSILES – VIETNAM**

Deployed firing batteries were issued six missiles for each launcher: three missiles on the launcher and another three stored on truck- or trailer-mounted missile pallets. Thus, firing batteries in square battalions had 36 missiles while batteries in triad battalions had 54. Missiles were delivered to the firing batteries in metal shipping containers. Upon receipt, unit personnel removed each missile from its container, inspected it for damage, attached the wings to the missile body, and then placed it on a missile pallet for storage until loaded onto a launcher. In this illustration, four crewmen are transferring missiles onto a launcher. The loader-transporter driver and the two other crewmen are taking directions from the officer standing next to the launcher. Because of the loader-transporter's engine noise, the officer is using hand signals to direct the crew. Reloading a launcher involved removing missiles from a trailer-mounted pallet (1), transporting the missiles to a launcher (2), and then transferring the missiles onto the launcher (3). A well-trained crew could transfer missiles from a storage pallet to a launcher in 15 minutes or less.

Damage to a missile was a serious matter. This missile on A Btry, 6-52 ADA's tactical site in 1981 fell from its launcher when part of the launcher's boom failed during system checks. (M. Romanych)

Aschaffenburg; however, France abruptly withdrew its battalions after leaving NATO. To cover this gap, 6-61 Arty stayed in Landshut until replaced by a German battalion in 1968, then redeployed back to Fort Bliss as a Return of Forces to Germany (REFORGER) unit designated for deployment to Germany if tensions increased.

US HAWK units were placed under the command of the 32d Artillery Brigade (redesignated in 1966 as 32d Army Air Defense Command, or AADCOM). The brigade was organized into four air defense groups (redesignated as ADA brigades in the early 1980s): 10th, 69th, 94th, and 108th. Initially, all HAWK battalions were assigned to either 10th or 69th Artillery Groups. Years later, 94th and 108th ADA Brigades each gained one battalion when two HAWK battalions moved from the HAWK Belt to the rear area for airbase defense.

Not only did the location of battalions shift during the 33 years US HAWK was stationed in Germany, but battalion organization and designations changed too. In the late 1960s, the four battalions closest to East Germany (3-7, 6-60, 6-517, and 6-562 Arty) were converted to self-propelled HAWK and given the dual role of defending the HAWK Belt and protecting US maneuver units. About the same time, 6-62 Arty left the Aschaffenburg area for former Nike missile sites west of the River Rhine near Spangdahlem and Bitburg airbases. In the early 1970s, six battalions were redesignated: 4-57 Arty became 2-57 ADA, 6-59 Arty was changed to 3-59 ADA, 6-60 Arty was reflagged 3-60 ADA, 6-62 Arty became 2-62 ADA, 6-517 Arty was redesignated 2-2 ADA, and 6-562 Arty became 1-1 ADA. More battalion location changes and designations occurred in the mid-to-late 1980s when 3-59 ADA moved from Hanau to Neubrücke to protect airbases in the rear area, and three battalions were redesignated: 1-1 ADA became 3-52 ADA, 2-62 ADA was reflagged as 1-1 ADA, and 3-59 ADA became 4-1 ADA.

Other European nations also acquired HAWK systems, extending NATO's air defense coverage into Northern and Southern Europe. In the mid-1960s, Denmark deployed four HAWK batteries around Copenhagen,

F

CONSTANT READINESS – GERMANY

This launcher crew on a tactical site in Germany during the mid-1980s is checking a launcher to ensure it is operational. Serving as front-line units, HAWK firing batteries maintained a constant state of readiness against a possible enemy air attack. When in 20-Minute or "Hot" alert status, a battery kept its HAWK system energized and operational, ready to fire a missile in 20 minutes or less. The HAWK crew manned the Battery Control Central, monitored the air picture provided by its radars, and maintained communications with the battalion fire direction center. The crew also checked the HAWK system every six hours to ensure it was fully operational. When in "Hot Status," crews typically worked shifts for 24 hours or longer. A typical "Hot Status" crew consisted of 20 operators, maintenance, and communications personnel. The average age of the crew was about 21 years old. Because of HAWK's grueling manning requirements, unit personnel often jokingly referred to HAWK as the "Holiday And Weekend Killer."

followed by another four batteries in 1983, closing the gap between Copenhagen and the northern end of the HAWK Belt in Germany. Meanwhile, Norway deployed six batteries to protect its airbases, and Sweden (not part of NATO) deployed four batteries around Stockholm. In Southern Europe, Italy stationed 16 batteries in northern Italy, and Greece positioned eight batteries – including some self-propelled launchers – around Athens and Thessaloniki. Greece was also host to the NATO Missile Firing Installation (NAMFI) in Crete, where Europe-based HAWK units live-fired missiles. In 1965, Spain placed three batteries near Gibraltar, later receiving another three batteries in 1990.

Even after withdrawing from NATO, France continued to operate 12 HAWK batteries for its own national defense. Notably, France was the only European nation to fire a HAWK missile in combat. In 1986, France deployed troops to Chad to counter a Libyan invasion, including a HAWK battery emplaced near the capital city of N'Djamena. On September 7, 1987, the battery shot down a Libyan Tu-22B bomber attempting to bomb the city's airport. Missile intercept occurred almost directly over the battery, and debris and unexploded bombs from the Tu-22 fell on the HAWK position, but no soldiers were injured or equipment damaged.

The Soviets designated HAWK batteries as high priority targets, employing espionage, ground and airborne signals intelligence, and aircraft flyovers to collect intelligence for disrupting HAWK unit operations. The Soviet Air Force also developed tactics to open corridors through the HAWK Belt by overwhelming firing batteries with formations of aircraft while suppressing HAWK PARs and HIPIRs with airborne jammers and anti-radiation missiles. To train their pilots for these attacks, the Soviets built mock HAWK sites on several air-to-ground ranges. However, NATO's HAWK battalions did not plan to fight from tactical sites unless defending against a surprise air attack, and other than building revetments to protect radars, most sites were not hardened significantly. Time permitting, battalions were expected to move to previously selected field locations, change radar frequencies, and institute radar emission control procedures to degrade Soviet intelligence. Yet the specter of being overrun by waves of Warsaw Pact aircraft remained. Even under ideal conditions, a HAWK battalion with 72 missiles on its launchers could destroy only 50–60 aircraft before being overwhelmed. This scenario led to the development of PIP III's LASHE system.

In the 1980s, US HAWK sites were targets of anti-war protests, incursions, vandalism, and terrorist activity. The most serious event was the bombing of D Btry, 3-59 ADA by the Red Army Faction (also known as the Baader-Meinhof Gang) in September 1985 which heavily damaged three radars. In response, 32d AADCOM increased HAWK site security by placing geese on tactical sites to

Soldiers of A Btry, 6-52 ADA emplace an AN/MPQ-55 Continuous Wave Radar during a field exercise in 1981. The soldier on the radar is unlocking the antenna so it can rotate freely once the radar is energized. The soldiers in the foreground are running the radar's power and data cables. (M. Romanych)

HAWK units based in the United States often underwent mobility exercises. This C-130 Hercules, at Fort Bragg in 1984, has just performed a rapid unloading of a Platoon Command Post and tracking radar from C Btry, 3-68 ADA while the airplane's engines were still running. (M. Romanych)

provide early warning of intrusions; however, because of logistical problems, this program was cancelled about a year later.

To make way for Patriot missile units, US HAWK battalions began deactivating in 1985, starting with 2-2 ADA in Giessen and 2-57 ADA in Ansbach, followed by 3-7 ADA in Schweinfurt, and then 3-60 ADA in Grafenwöhr. With these units inactivated, the HAWK Belt in southern Germany ceased to exist. After the Berlin Wall fell and the Soviet Army began withdrawing from Central Europe, HAWK and Patriot units were released from their 24-hour NATO air defense mission. Although plans called for several HAWK battalions to become corps air defense units, the remaining four battalions (1-1 ADA, 4-1 ADA, 3-52 ADA, and 6-52 ADA) inactivated in 1993. After 32 years, US HAWK's presence in Germany ended.

After the Soviets left Germany, other NATO countries slowly retired their HAWK systems: Belgium in 1994, Norway in 1998, Netherlands (which sold eight firing units to Romania) in 2004, Germany and Denmark in 2005, and Italy in 2011. Non-NATO countries Sweden and France kept their HAWK systems until 2010 and 2012, respectively. As of 2021, Greece, Spain, Sweden, Turkey, and Romania were still operating HAWK systems.

Middle East and the Gulf War

HAWK was widely deployed in the Middle East, where it conclusively proved its capabilities in combat. The first country in the region to deploy HAWK was Israel, which received its first Basic HAWK systems in 1965, and by the time of the 1967 Six Day War had eight operational firing batteries. On the first day of the war, in the first-ever combat firing of the HAWK system, a battery protecting the Dimona nuclear reactor downed a friendly Dassault MD.450 Ouragan fighter-bomber that flew too close to the reactor. During the so-called "War of Attrition" (1967–70), Israeli HAWK destroyed 15 Egyptian aircraft, including MiG-17 and MiG-21 fighters, Su-7 fighter-bombers, and an IL-28 bomber. Israeli HAWK shot down even more enemy aircraft during the October 1973 War. At the outset of the war, Israel had 12 HAWK batteries, including some with self-propelled launchers. Egyptian and Syrian aircraft aggressively attacked several HAWK sites, damaging or destroying at least

An AN/MPQ-57 HIPIR of C Btry, 3-68 ADA being mated to its trailer after being off-loaded from a C-130 transport aircraft during a mobility exercise at Fort Bragg in 1984. (M. Romanych)

two missile systems, yet, Israeli HAWK batteries still shot down 22 aircraft. Lastly, during the 1982 Lebanon War, an Israeli HAWK battery downed a Syrian MiG-25 flying at high altitude near the Lebanon border. In all, between 1967 and 1982, Israeli HAWK batteries shot down as many as 38 enemy aircraft. Having proved itself so well in combat, the HAWK system remained a mainstay of Israeli air defense forces until the mid-2010s.

Impressed with Israel's success, many other countries in the Middle East also purchased HAWK systems. In the 1970s, Iran purchased 24 HAWK systems, Kuwait four, and Jordan 15; and in the 1980s, the United Arab Emirates purchased six systems, Saudi Arabia 16, and Egypt 12. Of these countries, only two – Iran and Kuwait – fired HAWK missiles in combat. Iran, despite having lost US logistical support after the 1979 Revolution, employed HAWK systems during the Iran–Iraq War, claiming to have downed at least 40 Iraqi aircraft, including Su-22s, MiG-23s, and

G

MOBILITY – WHITE SANDS MISSILE RANGE

Quickly emplacing and bringing the HAWK system into action was a well-choreographed event. First, the radars, control vans, and launchers were towed into place and disconnected from their prime movers. Then, operators and maintenance personnel, organized into crews of three or four soldiers to each piece of equipment, leveled and electrically grounded the radars, control vans, and launchers. The crews then ran data and power cables to link the radars and launchers to the control vans and to draw power from the diesel generators. Once cables were connected, each piece of equipment was energized and function checked. Meanwhile, the radars, Identification Friend or Foe system, and launchers were aligned to north, and missiles were loaded onto the launchers. The last step was to function check the entire HAWK system to ensure it could detect, track, and engage targets.

This illustration shows soldiers emplacing a High Power Illuminator Radar. The soldier on the radar is aligning the radar to the rest of the HAWK system. Scattered around the base of the radar are cable reels, camouflage poles and netting, other equipment needed to operate the radar. Spare lengths of cable are laid in a figure-eight configuration to prevent them from overheating (a coiled power cable would overheat because of the magnetic field produced by electricity passing through the cable).

a MiG-25. Reportedly, these same HAWK units shot down three Iranian F-14s and an F-4. Iraqi pilots referred to Iranian HAWK air defenses as "Death Valley" and employed jamming, anti-radiation missiles, and air strikes against HAWK sites, or more often, simply avoided flying within missile range. Although Iraqi air attacks damaged or destroyed several HAWK batteries and a shortage of repair parts limited equipment readiness, the Iranian military still considered HAWK to be far more effective than any of its Soviet-built surface-to-air missile systems. After the war ended in 1988, Iran managed to keep several HAWK fire units operational by manufacturing its own parts and components for the next three decades, until it unveiled the Mersad missile system, a derivative of HAWK with modified or new radars and missiles in 2010.

The other Middle Eastern country to use the HAWK system in combat was Kuwait, which purchased four systems in 1975. During the Iraqi invasion of Kuwait in 1990, two HAWK batteries operating near Kuwait City and a firing platoon emplaced on an island in the Persian Gulf reportedly downed 23 Iraqi aircraft, including MiG-23 and Su-22 jets, and a helicopter. After capturing the Kuwaiti HAWK systems, the Iraqi military emplaced at least one system near Baghdad, but was unable to operate it. Mindful of the potential threat posed by the system, the US Air Force reprogrammed its aircraft jammers and radar warning receivers before *Desert Storm* to counter HAWK radars.

Less than two weeks after Iraq occupied Kuwait, US Army and Marine Corps HAWK units were sent to the Persian Gulf to defend points of entry and assembly areas. The 2d LAAM Bn from Yuma arrived by air on August 14, 1990, and was placed under command of Marine Air Control Squadron 1 to protect Shaikh Isa Air Base in Bahrain and King Abdul Aziz Naval Base in Saudi Arabia. In October, Army Patriot and HAWK units arrived and were organized into HAWK–Patriot Taskforces (TF) under 11th ADA Brigade. Taskforce Scorpion (with Patriot batteries from 3-43 ADA and HAWK batteries from 2-1 ADA) arrived from Fort Bliss to support the XVIII Airborne Corps, and TF 8-43 ADA (with Patriot batteries from 8-43 ADA and HAWK batteries from 6-52 ADA) arrived from Germany with VII Corps. More HAWK batteries arrived in January. The 3d LAAM Bn

In the 1980s, HAWK units often participated in major deployment exercises to demonstrate American military resolve and the ability to reinforce its allies worldwide. Camouflage netting conceals this AN/MPQ-57 HIPIR during a Bright Star exercise in Egypt in 1985. (NARA)

arrived by ship from Cherry Point and relieved the 2d LAAM Bn, which then moved forward to the Kuwait border, and Army HAWK batteries from 2-52 ADA arrived from Fort Bragg and were attached to 2-43 ADA (Patriot) from Germany to protect the King Khalid Military City logistics base. Lastly, HAWK batteries from Saudi Arabia and the United Arab Emirates were employed to defend rear area assets while Netherlands deployed two HAWK batteries and Germany one battery as part of a HAWK–Patriot taskforce from Europe to protect the airbase at Diyarbakir, Turkey.

On paper, the Iraqi Air Force had more than 700 combat aircraft, including modern MiG-29, Su-25, and Mirage F1 airplanes, and possessed airborne jammers and anti-radiation missiles built by the Soviets for the sole purpose of attacking HAWK HIPIRs. During *Desert Shield*, most HAWK firing units were positioned on the forward edge of the Coalition's air defense zone, which was occasionally probed by Iraqi warplanes and helicopters. To conceal their locations and mitigate the threat of anti-radiation missiles, the forward-most HAWK units limited radar emissions, relying on external air picture and target data from other sources such as the E-3 AWACS. After the air campaign commenced on January 17, 1991, the Coalition quickly achieved air superiority. Iraqi aircraft continued to probe Coalition air defenses, but none entered Coalition airspace, and no HAWK missiles were fired. The threat of an air attack subsided when much of the Iraqi Air Force fled to Iran on January 26. HAWK and Patriot units remained on alert, and TF Scorpion, TF 8-43, and 2d LAAM Bn moved forward to the Iraq and Kuwait borders to provide air defense coverage of the Tapline Road (the major highway along the northern border of Saudi Arabia), unit assembly areas, and border breach points. To help mask this movement, VII Corps' deception detachment used electronic transmitters to fake HAWK radar emissions. On the second day of the ground offensive, 2-1 ADA's B Btry and 6-52 ADA's A and C Btrys moved into Iraq to extend coverage over forward units of the XVIII Airborne and VII Corps. Two days later, 2d LAAM Bn moved into southern Kuwait with the 1st Marine Division. Several HAWK batteries were operational in Iraq when fighting ceased on February 27. When Coalition air defense operations were terminated in early March, the Army and Marine Corps redeployed the HAWK battalions back to their home stations.

After the Gulf War, Kuwait did not replace its HAWK systems; however, concerned about regional stability, nearby Bahrain purchased six PIP III systems and Turkey purchased 12 HAWK XXI systems. In 2020 Turkey deployed several of its systems along its border with Syria to deter Syrian and Russian airstrikes, and to Libya to protect UN-recognized government forces from air attack by the rival Libya National Army (LNA).

US-based battalions

From their inception, the Marine Corps' HAWK battalions were organized and trained specifically for expeditionary operations. In this role, 3d LAAM Bn, stationed at Cherry Point, deployed a firing battery to Cuba in 1962, and 1st and 2d LAAM Bns, stationed at Twentynine Palms, deployed to Vietnam in 1965. After the war, the Marine Corps reduced its HAWK battalions to three: 2d LAAM Bn at Yuma, 3d LAAM Bn at Cherry Point, and the reserve force 4th LAAM Bn in California. When the Cold War heated up, 1st LAAM Bn was reactivated and stationed on Okinawa from 1987–90 to support the

Two launchers of C Btry, 2-1 ADA emplaced somewhere in northern Saudi Arabia during Operation *Desert Shield* in January 1992. Although no HAWK units fired on Iraqi aircraft during the Gulf War, their presence deterred Iraqi air strikes and aerial surveillance. (NARA)

III MAF in the Pacific Area of Operations. Deployment of the 2d and 3d LAAM Bns to the Gulf War was Marine HAWK's last mission before being retired in 1997.

From 1962 to 1994, several Army HAWK battalions were stationed at Fort Bliss as deployable contingency units. When not preparing for contingency missions, these battalions frequently participated in strategic deployment exercises and supported research and development programs. Battalions stationed at Fort Bliss during that time were: 8-7 Arty (later reflagged as 1-7 ADA and then again as 2-1 ADA) from 1962–94, 6-56 and 6-71 Arty before and after their deployment to Vietnam, 6-61 Arty (later reflagged as 2-55 ADA) from 1968–85, and 1-65 Arty, which returned to Fort Bliss from Key West in 1979 and was later reflagged as 3-1 ADA and sent to Fort Hood in 1988. To maintain preparedness of the deployable battalions, the 11th Artillery Group (later redesignated as 11th ADA Brigade) was activated in 1971. The brigade headquarters deployed during the Gulf War to provide tactical control of Army HAWK and Patriot units.

In the late 1970s, the Army began assigning HAWK battalions to serve as organic air defense to its three US-based corps. In 1978, I Corps at Fort Lewis received the newly formed 1-4 ADA (later redesignated 1-52 ADA). The XVIII Airborne Corps at Fort Bragg received 3-68 ADA (later redesignated to 2-52 ADA) after the battalion was released from its mission at Homestead Air Force Base in 1979, while III Corps received 3-1 ADA, which was transferred from Fort Bliss to Fort Hood in 1988. In addition, 2-51 ADA was activated in 1978 and stationed at Fort Riley, Kansas with the 1st Infantry Division for only seven years. After the Gulf War, all active Army HAWK battalions (including those in the US) were inactivated by 1994. However, HAWK equipment continued to be used on electronic warfare training ranges such as Nellis Air Force Base, Nevada well into the 21st century.

As the Army removed its HAWK units from active duty, the Army National Guard formed three HAWK battalions (2-174 ADA in Ohio, 7-200 ADA in New Mexico, and 2-265 ADA in Florida) as replacements. An additional two battalions (1-263 ADA in South Carolina and 2-200 ADA in New Mexico) were not completely formed before the Army decided to remove all HAWK systems from service. Unable to sustain its HAWK systems on its own, the

National Guard either inactivated or converted its HAWK battalions to Avenger air defense units, and by 1997 the last battalion was gone.

LEGACY OF THE HAWK MISSILE SYSTEM

For more than 30 years, the HAWK missile system was the mainstay of US Army and Marine Corps air defense artillery and a crucial part of the US and NATO arsenals. HAWK was also the most widely used non-Soviet surface-to-air missile system during the Cold War. Although neither the US Army nor Marine Corps fired a HAWK missile at a hostile aircraft, HAWK was a success as a defensive weapon system, demonstrating US resolve and deterring air attack by the Soviet Union and its allies. In the hands of other militaries, the HAWK system proved lethal when used in combat, readily destroying even the most modern and sophisticated aircraft of the Cold War.

Today, HAWK remains in service with many nations – Singapore, Egypt, Turkey, Bahrain, Jordan, Greece, Morocco, Saudi Arabia, Spain, Romania, and the United Arab Emirates. A forerunner of 21st-century surface-to-air missile systems, many of HAWK's design features were incorporated into the next generation of air defense systems such as Norway's NASAMS, Israel's David's Sling, Iran's Mersad, Sweden's Robot system 97, and Taiwan's Sky Bow, keeping the legacy of HAWK alive more than 60 years after it was first fielded.

BIBLIOGRAPHY

The HAWK missile system has been overlooked and undervalued by historians. Books and articles that mention HAWK provide only cursory descriptions of the system and its employment. The authors of this book consulted a wide range of primary sources, including field and technical manuals, official records and unit reports, and interviews of military and civilians who worked with the HAWK system. The periodicals *US Army Air Defense Digest*, *Air Defense Trends*, and *Air Defense Artillery* published by the US Army Air Defense School were especially useful, as were the following publications:

Crabtree, James D., *On Air Defense*, Praeger, Westport, CT (1994)

Crabtree, James D., *Secrets and Scuds: An Untold Story of Desert Shield and Desert Storm*, Jimmy Sinai Books, Fayetteville, NC (2016)

Hamilton, John A., *Blazing Skies: Air Defense Artillery on Fort Bliss, Texas, 1940–2009*, Department of the Army, Arlington, VA (2009)

McKenny, Janice E., *Air Defense Artillery*, Center of Military History, US Army, Fort Lesley J. McNair, Washington DC (1985)

FM 44-90, *Air Defense Artillery Employment, HAWK*, Department of the Army, Arlington, VA (1977)

FM 44-90, *HAWK Battalion Operations*, Department of the Army, Arlington, VA (1987)

FM 44-96, *Air Defense Artillery Missile Unit, HAWK (Battalion and Battery)*, Department of the Army, Arlington, VA (1965)

FM 44-96, *Air Defense Artillery Employment, HAWK*, Department of the Army, Arlington, VA (1971)

INDEX

Note: locators in bold refer to plates, illustrations and captions.